A WORLD IN TRANSITION; HUMANKIND AND NATURE

AF788348

EINSTEIN MEETS MAGRITTE: An Interdisciplinary Reflection on Science, Nature, Art, Human Action and Society

Series Editor

Diederik Aerts, Center Leo Apostel, Vrije Universiteit Brussel, Belgium

---

Volume 1 *Einstein Meets Magritte: An Interdisciplinary Reflection*
The White Book of 'Einstein Meets Magritte'
Edited by Diederik Aerts, Jan Broekaert and Ernest Mathijs

Volume 2 *Science and Art*
The Red Book of 'Einstein Meets Magritte'
Edited by Diederik Aerts, Ernest Mathijs and Bert Mosselmans

Volume 3 *Science, Technology, and Social Change*
The Orange Book of 'Einstein Meets Magritte'
Edited by Diederik Aerts, Serge Gutwirth, Sonja Smets and Luk Van Langenhove

Volume 4 *World Views and the Problem of Synthesis*
The Yellow Book of 'Einstein Meets Magritte'
Edited by Diederik Aerts, Hubert Van Belle and Jan Van der Veken

Volume 5 *A World in Transition; Humankind and Nature*
The Green Book of 'Einstein Meets Magritte'
Edited by Diederik Aerts, Jan Broekaert and Willy Weyns

Volume 6 *Metadebates on Science*
The Blue Book of 'Einstein Meets Magritte'
Edited by Gustaaf C. Cornelis, Sonja Smets, Jean Paul Van Bendegem

Volume 7 *Quantum Structures and the Nature of Reality*
The Indigo Book of 'Einstein Meets Magritte'
Edited by Diederik Aerts and Jarosław Pykacz

Volume 8 *The Evolution of Complexity*
The Violet Book of 'Einstein Meets Magritte'
Edited by Francis Heylighen, Johan Bollen and Alexander Riegler

---

VOLUME 5

A World In Transition; Humankind and Nature

The Green Book of "Einstein Meets Magritte"

Edited by

Diederik Aerts, Jan Broekaert and Willy Weyns
Brussels Free University

VUB UNIVERSITY PRESS
VRIJE UNIVERSITEIT BRUSSEL
BELGIUM

SPRINGER SCIENCE+BUSINESS MEDIA, LLC

Library of congress Cataloging-in-Publication Data

ISBN 978-94-010-3741-9 ISBN 978-94-007-0856-3 (eBook)
DOI 10.1007/978-94-007-0856-3

Printed on acid-free paper

All Rights Reserved
© 1999 Springer Science+Business Media New York
Ursprünglich erschienen bei Kluwer Academic Publishers and Vrije Universiteit Brussel 1999
Softcover reprint of the hardcover 1st edition 1999
No part of the material protected by this copyright notice may be reproduced or
utilized in any form or by any means, electronic or mechanical,
including photocopying, recording or by any information storage and
retrieval system, without written permission from the copyright owner.

Table of contents

DIEDERIK AERTS

THE GENERAL INTRODUCTION OF *EINSTEIN MEETS MAGRITTE*

The series of books 'Einstein meets Magritte' presented here originates from an international interdisciplinary conference with the same title, which took place in Brussels in spring 1995. On the eve of the third millennium, we assembled scientists and artists to reflect together on the deep nature of reality and the knowledge and skill humankind has gathered in this field. We had decided to call this meeting 'Einstein meets Magritte' because we believed that meaningful keys could be found at the place where the two meet. It is the way of the world that has made Einstein and Magritte into icons of our culture. The purpose of the conference was to reflect and debate without fear on the most profound and timeless questions.

On one of those evenings, when the talks and discussions were long and exhausting and the press were doing all they could to get Albert Einstein and René Magritte in front of the microphones and cameras, a few of my most loyal aides and myself succeeded in getting them safely and quietly to a taxi, which then carried us off into the Brussels night. We got out at Manneken Pis, since that was on Einstein's list, and we concealed ourselves among the many tourists who were coming and going, expressing their wonder in every language under the sun at the famous little statue. And one of us was taking pictures: Einstein and Magritte leaning against the railings, with us beside them, and one more, arm in arm, and then another in case the first was no good, when suddenly I felt a heavy slap on my shoulder:

"How you doing, mate?"

It was Jacky and his inseparable girlfriends Nicole and Sylvie, and everyone embraced everyone else. I introduced Albert and René, and interest was immediately shown, and I had my heart in my mouth, because Jacky was a painter, poet and urban philosopher. We walked together through the alleys of Brussels in dismal Belgian rain, over cobblestones that glistened in the street lamps.

When we had provided for the inner man with 'Rabbit in Beer' and 'Mussels with Fries', and finally a 'Dame Blanche' topped with warm chocolate sauce as apotheosis, Jacky enticed us to his house in the Rue Haute where we threw ourselves into deep, soft armchairs. Albert and René were offered the best places and as always Jacky told the story

© 1999 *Kluwer Academic Publishers. Printed in The Netherlands.*

of his life and discussed his rightness, as he did repeatedly, with a confidence and suppleness that distinguished him so sharply from modern science. Albert listened enthralled and René was fascinated, and once more my heart was in my mouth, but Nicole winked reassuringly, and Sylvie brought us snacks on cushions of Brussels lace and sweet white wine in tall, old-fashioned crystal glasses.

The topic of discussion for the evening turned out to be 'the doubts of modern science'. In science there is not a single hypothesis for which one cannot find two groups of hard-working scientists, one of which can 'prove' the hypothesis while the other can 'prove' its negation. And the more fundamental and important the question is, the more clearly the situation turns out like this.

"It's crazy," maintained Jacky, "In fact science states that one doesn't know anything anymore."

"That's right," said Albert, "Truth is not a simple concept, and I believe that the history of science makes it clear how often erroneous hypotheses have been believed over the centuries."

"A good thing too," replied René, "Things can only happen as a result of the movement brought about by that constant doubt."

Meanwhile Sylvie came to join us and handed round pictures of the exhibitions of Jacky's paintings and poems. Jacky suddenly got very excited, as if something had inspired him, and he leapt up and vanished into his studio. A few minutes later he returned with his palette and brush poised. Before I could stop him he had started painting violently right at the spot where Albert and René were sitting. A large, gossamer-thin piece of Brussels lace gradually took shape and Albert and René vanished. Fortunately, my young assistants, Jacky's girlfriends and myself got away with just a few vicious daubs of paint in the face.

The series of eight volumes introduced here are not just the results of the conference, as would be the case with a record of the proceedings. The authors were invited to write with the events at the conference in the back of their mind, so that the books would form a second phase in the process of thought set in motion at the conference. A second phase more clearly crystallised than the self-organising forum that arose during the conference, but one which focuses on the same timeless questions and problems.

The whole ensemble was already streamlined at the conference into a number of main topics named after the colours of the rainbow - red, orange, yellow, green, blue, indigo and violet, as well as white, the synthesis of all colours. This order was maintained and led to eight separate books in the series.

Volume 1: Science, Nature, Human Action and Society: an Interdisciplinary Reflection: the White Book of *Einstein meets Magritte*

The white book contains more fully developed versions of the contributions made by the keynote speakers at the conference. So this white book covers various scientific topics. In his article, 'Basically, it's purely academic', John Ziman asks himself what 'basic research' really is in today's world. In his contribution, 'The manifest image and the scientific image', Bas Van Fraassen analyses the considerable differences between the theoretical scientific description of the world and the way it appears to us. He argues that most formulations of this problem may themselves be tendentious metaphysics, full of false contrasts, and that insistence on a radical separation between science and what we have apart from science, and on the impossibility of accommodating science without surrender, may be a way of either idolising or demonising science rather than understanding it. In the 'Microdynamics of incommensurability: philosophy of science meets science studies', Barbara Hernnstein-Smith examines the bemusing but instructive logical, rhetorical and cognitive dynamics of contemporary theoretical controversy about science. In his contribution 'Subjects, objects, data and values', Robert Pirsig proposes a radical integration of science and value that does no harm to either. It is argued that values can exist as a part of scientific data, but outside any subject or object. This argument opens a door to a 'metaphysics of value' that provides a fundamentally different but not unscientific way of understanding the world. Ilya Prigogine discusses in 'Einstein and Magritte: a study of creativity', the global transformation of a classical science which was based on certainties into a new science that takes possibilities as its basic concepts. Constantin Piron demonstrates in his contribution 'Quanta and relativity: two failed revolutions' that none of the two great revolutions in physics, quantum mechanics and relativity theory, have actually been digested by the physics community. He claims that the vast majority of physicists still cling to the idea of a non-existent void full of little particles, in the spirit of Leibniz or Descartes. Rom Harré reflects on the significance of the theory of relativity. In his article 'The redundancy of spacetime: relativity from Cusa to Einstein', he defends the hypothesis that relativity theory is best interpreted as a grammar for coordinating narratives told by different observers. In his contribution 'The stuff the world is made of: physics and reality', Diederik Aerts analyses the consequences of the recent advances in quantum mechanics, theoretically as well as experimentally, for the nature of reality. He analyses the deep conceptual paradoxes in the light of these recent data and tries to picture a coherent model of the world. In his contribution 'Dasein's brain: phenomenology meets cognitive science', Francisco Varela puts forward the hypothesis that the relation between brain processes and living human experience is the really hard problem of consciousness. He argues that science needs to be complemented by a deep scientific investigation of experience itself to move this major question beyond the sterile oppositions of dualism and reductionism. In his contribution 'What creativity in art and science tell us about how the brain must work' William Calvin defends the prospects for a mental Darwinism that operates on the milliseconds to minutes time scale, forming novel ideas and sentences never previously expressed. Adolf Grünbaum in his article 'The hermeneutic versus the scientific conception of psychoanalysis: an unsuccessful effort to chart a via media for the human sciences' argues that the so called 'hermeneutic' reconstruction of psychoanalytic theory and therapy proposed by Karl Jaspers, Paul Ricoeur and Jürgen Habermas fails both as a channel and as alleged prototype for the study of human nature. In his article 'Immortality, biology and computers', Zygmunt Bauman analyses the shift that postmodern society has provoked regarding the concept of immortality. He points out that strategies of collective and individual immortality have shifted from the modern deconstruction of death to a postmodern deconstruction of immortality, and points out that the possible consequences of this process need to be taken into consideration. Brian Arthur, in his article 'The end of certainty in economics', points out that our economy is very non-classical, meaning that it is based on essentially self-referential systems of beliefs about future economic conditions. He argues that our economy is inherently complex, subjective, ever-changing, and to an unavoidable degree ill-defined.

Volume 2: Science and Art: the Red Book of *Einstein meets Magritte*

And then Magritte comes in. Many obvious differences exist between science and art. But the *Science and Art* volume of this series addresses not only these differences but also the possibilities of crossing several of the gaps between science and art. Several contributions deal with sociological and philosophical elaborations of the similarities and differences between science and art, while others approach science from an artistic point of view and art from a scientific point of view. The volume also considers several approaches that attempt to go beyond the classical dichotomy between the two activities. In a special section, attention is paid to the particular role played by perception in both science and art as a regulator of human understanding. Together, these contributions strive for an intensive interaction between science and art, and to a consideration of them as converging rather than diverging. It is to be hoped that both science and art will benefit from this attempt.

Volume 3: Science, Technology and Social Change: the Orange Book of *Einstein meets Magritte*

The major subject of the orange book is that society as a whole is changing, due to changes in technology, economy and the changing strategies and discourses of social scientists. The collected articles in the orange stream discuss a range of specific societal problems related to the subject of social change, the topics of the articles range from the scale of for instance sociology of health and psychohistory to more specific social problems like for instance anorexia nervosa, art academies and the information superhighway. Although the authors approach different subject matters from dissimilar perspectives and work with various methods, all the papers are related to the theme of science, technology and social change. In the orange book the reader will find a lot of arguments and hints pertaining to questions like: To what exactly will this social change lead in the 21st century? What kind of society lies ahead? She/he will be confronted to a plethora of enriching conceptions of the relationships between social sciences and social changes.

Volume 4: World Views and the Problem of Synthesis: the Yellow Book of *Einstein meets Magritte*

A rapidly evolving world is seen to entail ideological, social, political, cultural and scientific fragmentation. Many cultures, subcultures and cultural fragments state their views assertively, while science progresses in increasingly narrowly defined areas of inquiry, widening not only the chasm between specialists and the layman, but also preventing specialists from having an overall view of their discipline. What are the motive forces behind this process of fragmentation, what are its effects? Are they truly inhospitable to the idea of synthesis, or do they call out, more urgently than ever before, for new forms of synthesis? What conditions would have to be met by contemporary synthesis? These and related questions will be addressed in the yellow book of Einstein meets Magritte.

Volume 5: A World in Transition; Humankind and Nature: the Green Book of *Einstein meets Magritte*

'A World in Transition. Humankind and Nature' is appropriately entitled after its aim for an intrinsic property of reality: change. Of major concern, in this era of transformation, is the extensive and profound interaction of humankind with nature. The global scaled, social and technological project of humankind definitely involves a myriad of changes of

the ecosphere. This book develops, from the call for an interdisciplinary synthesis and respect for plurality, acknowledging the evolving scientific truth, the need for an integrated but inevitably provisional world view. Contributors from different parts of the world focus on four modes of change: i) Social change and the individual condition, ii) Complex evolution and fundamental emergent transformations, iii) Ecological transformation and responsibility inquiries, iv) The economic-ecological and socio-technical equilibria. Primarily reflecting on the deep transformations of humankind and on the relationship between humans and nature it addresses major points of contemporary concern.

Volume 6: Metadebates: The Blue Book of *Einstein meets Magritte*

This book provides a meta-disciplinary reflection on science, nature, human action and society. It pertains to a dialogue between scientists, sociologists of science, historians and philosophers of science. It covers several topics: (1) the relation between science and philosophy, (2) new approaches to cognitive science, (3) reflections on classical thinking and contemporary science, (4) empirical epistemology, (5) epistemology of quantum mechanics. Indeed, quantum mechanics is a discipline which deserves and receives special attention here, for it still is a fascinating and intriguing discipline from a historiographical and philosophical point of view. This book does not only contain articles on a general level, it also provides new insights and bold, even provocative theories on the meta-level. That way, the reader gets acquainted with 'science in the making', sitting in the front row.

Volume 7: Quantum Structures and the Nature of Reality: The Indigo Book of *Einstein meets Magritte*

This book refers to the satellite symposium that was organised by the International Quantum Structure Association (IQSA) at Einstein meets Magritte. The IQSA is a society for the advancement and dissemination of theories about structures based on quantum mechanics in their physical, mathematical, philosophical, applied and interdisciplinary aspects. The book contains several contributions presenting different fields of research in quantum structures. A great effort has been made to present some of the more technical aspects of quantum structures for a wide audience. Some parts of the articles are explanatory, sketching the historical development of research into quantum structures, while other parts make an effort to analyse the way the study of quantum structures has contributed to an understanding of the nature of our reality.

Volume 8: The Evolution of Complexity: The Violet Book of *Einstein meets Magritte*

The violet book collects the contributions that consider theories of evolution and self-organisation, on the one hand, and systems theory and cybernetics, on the other hand. Both can add to the development of an integrated world view. The basic idea is that evolution leads to the spontaneous emergence of systems of higher and higher complexity or "intelligence": from elementary particles, via atoms, molecules, living cells, multicellular organisms, plants, and animals to human beings, culture and society. This perspective makes it possible to unify knowledge from presently separate disciplines: physics, chemistry, biology, psychology, sociology, etc. The volume thus wishes to revive the transdisciplinary tradition of general systems theory by integrating the recently developed insights of the "complex adaptive systems" approach, pioneered among others by the Santa Fe Institute.

Even these books only signify a single phase in the ever-recurring process of thought and creation regarding the basic questions on the reality that surrounds us and our place in it.

Brussels,
July 17, 1998.

JAN BROEKAERT, WILLY WEYNS

A WORLD IN TRANSITION; HUMANKIND AND NATURE

EDITORIAL

Change is an intrinsic property of reality, whether slowly evolving or rapidly transforming. The concomitant re-identification of concepts and redefinition of relations between entities or systems is a continuous task and intellectual and emotional challenge. Not only is humankind, driven to identification of origins and causes, inquiring into mechanisms and deriving prediction of consequences of change, but also inclined to evaluation of evolutions and action against or in approval of change. A contemporary and obvious observation of our world reveals the realm of change is spreading in fields as many and as different as; nature, politics, labor, health, spirituality, education, culture and technology.

Changes can be welcomed and induced, but as well, unwanted and without identifiable origins or without definite causality, ultimately beyond mankind's control. Of major concern, in this era of transformation, is the extensive and profound interaction of humankind with nature. The global scaled, social and technological project of humankind definitely involves a myriad of changes of the ecosphere. Sometimes limited in impact and extent but nowadays some are of similar global extent and allegedly of the catastrophical kind. Whether humankind's evolution is responsible for, or whether nature is self-steering its evolution, the indissoluble unity of both spheres is the reality of the world of change. The omnipresence and complexity of the realm of change urges a broad spectrum of approach, such as the weighed complementarity of the configuration 'Einstein meets Magritte' conveys.

In the present volume, from the call for a necessarily interdisciplinary approach, and from the aim for synthesis and respect for plurality, acknowledging the evolving scientific truth, the need for an integrated but inevitably provisional world view and paradigm emerges. From within a coherent world view—albeit the plural 'views'—responsibility and engagement can be taken up in favor of a sustainably evolving whole of humankind and nature.

The contributions tentatively have been regrouped according to similarity of subject or domain of research. By no means this subdivision is necessary, it could to the same extent have been fruitfully clustered in many different ways. The applied subdivision of contributions proposes four modes of change:

© 1999 *Kluwer Academic Publishers. Printed in The Netherlands.*

1. **Social change and the individual condition.**
   From a mainly economical stance **Riccardo Petrella** describes the evolving faces of the capitalist paradigm and the need for future redirection of global policies through new world organisations. De-emphasising competitivity and tools leads to rethinking the social global future.
   In a systemic approach the barriers of cognitive evolution are investigated by **Borisz Szántó**. Not only the contradictory interrelatedness of the self-modifying cognitive progress of the individual—through education and emphasis on socio-technological possibilities—and the self-restoring but limited stability of the ecosphere, but also the linearity of an uncontrolled socio-technical environment sets boundaries to humankind's evolution.
   A mainly regional Russian analysis of the social transition crisis during the post Cold-War period is made by **Dmitri Spivak**. 'Apathy', 'fear' and 'mysticism' are recognised as key concepts for transposing altered states of consciousness to socio-cultural groups.
   A futurist speculation on emergent social freedom is drawn from insights in the ontology of quantum-mechanics and evolution by **Diederik Aerts**. Presented as a narrative account reminiscent to the Platonic dialogues, the individual's movement through social space and his emergent collaborations lead to a vision of an expanded social and interactive paradigm.
   The subtle mechanism of the individual's ultimate contestation of social injustice is evocatively unfolded by **Kyra Marie Landzelius**. The mediatised image of the starving body as an icon of absolute ethics results from self-destructive hunger strike.
2. **Complex evolution and fundamental emergent transformations.**
   In a thought experiment **Fenton F. Robb** explains how 'institutions' spontaneously emerge from social interaction as autopoietic and self-defining real-world entities. As supra-human 'institutes' define the human perception and understanding of the real, true, good and causally related, the need for research on this evolutionary change through 'Institutional Dynamics' seems necessary.
   A major problem in transdisciplinary studies is the exchange of terminology and process mechanisms between disciplines, conveyed by analogies and metaphors. **Czesław Mesjasz** analyses the effects of the introduction of system analogies and metaphors in the study of contemporary world politics. Not only is the metaphor's normative value enhanced, but as social systems are self-referential and autopoietic they contribute to their proper development.

Dramatic changes in society, technology and environment enhance the emergence of pathogenic agents, increase the complexity of health profiles and affect the predictability of dominant health transitions. According to **Marilia Bernardes Marques** the concomitant intrinsic complexity of health policy assessment urges a transdisciplinary and technological prospective approach.

A careful analysis of the emergence of 'significant universals' in language and music is exposed by **Philip L. Peterson**. Endowed with explanatory capacity and biologically unreducible these cultural entities are considered to exist distinctly from nature.

A speculative symbolic formalism to implement the idea of communication in biology is advanced by **Arnold De Loof**. The analysis, based on non-linear dynamical systems, amounts to the claim that 'communication between hierarchical constituents' must be considered essential in the concept of 'life'.

3. **Ecological transformation and responsibility inquiries.**
The subjective construction of scientific truth and the need for lawful mediation between concomitant power and resistance is approached theoretically by **Serge Gutwirth**. The privileged role of science in environmental regulations now submitted to truth-pluralism and law mediation could—contrary to its previously alleged destructive role—be democratically guided.
By means of a SWOT-analysis **Willy Weyns** describes positive and negative potentialities of humankind's techno-ecological relations. The break down of successful micro-level strategies when transposed to the global scale, calls for new solutions and revision of the contemporary paradigm.
From a spiritual stance **Jan Fennema** builds up a reflection on the fundamental integration of humans in nature. Deepened ecological knowledge or Wisdom—more than Promethean arrogance—is needed to avoid further environmental deterioration.
Peng Fu-Yang presents a moralist analysis of the alienating influence of over-emphasised science and technology development. Support is given to policies for co-evolution of humankind and nature. Paradoxically this is proposed by means of scientific achievement and at risk of sacrifying the individual integrity.

4. **The economic-ecological and socio-technical equilibria.**
Prior to the industrially and commercially induced deterioration of the environment, 'Indigenous Knowledge' achieved a sustainable and harmonious relation to the ecosystem. **Francis D. O'Reilly** now advises 'necessary knowledge' to strategically counter the glob-

alised economical pressure on Third World countries.
H Thomas R. Mark describes how the economic and ecological global system are both systems of hierarchically nested processes. The decoupling of both, causing depletion and deterioration of the environment, could be reversed through monetary institutions enforcing appropriate spatio-temporality on the global process.
The paradigm of sustainable development has been bypassed by the information society paradigm, ever relaunched by innovative technological achievement. **Riccardo Petrella** traces back the origins of the presently dominating paradigm to the capitalist industrial society paradigm. Applied differently, information and communication technology could facilitate the advent of a new global social contract.
In a historical survey of economy **Janina Rosicka** describes how the myth of fertility evolved into the opposed concepts of Scarcity and Plenty. The former became the axis of modern economy while the latter, reflected in the great social project, amounted to the concept of equality albeit on substantially low level.

Clearly all contributions are primarily concerned with the deep transformations of humankind and of the relationship between humans and nature. As a matter of fact, most contributions to the green book share what we are inclined to call a 'broad ecological vision' in their diagnostic of an increasing imbalance between humankind and its environment. Fortunately, as the reader going through this record will soon discover, this well documented factual statement is not their last word. The transition from the 'is' to the 'ought', from 'pure science' to ethics is generally made without hesitation. Almost all authors derive from their scientific description of the illness at least an indication of a remedy, each in their specific field of interest or discipline. No doubt, the multi-disciplinary background of the scientific contributors is one of the strong points and the originality of this truly international volume. The contributions presented here are from specialists in health research, biology, scientific ecology, philosophy, anthropology, sociology, physics, economy and information technology. Hence it is no surprise that the diagnoses of the ecological predicament and social transformations formulated by those distinguished contributors is extremely diverse, at least in first approach. Precisely these wide-ranging approaches allow creative associations between different contributions.

Some of the authors point at the human's belief of the myth of nature, and blame the 'Promethean arrogance' of the human species as a primal cause for the current state of affairs (Weyns, Fennema). Sometimes even intervention is proposed at the basic human level, putting first ecological integrity at the expense of individual integrity (Fu-Yang). Instrumentalism, defined as the precedence of the scientifico-technical realm over ethics is accused (Petrella, Mark, O'Reilly). Generally accepted concepts are not any longer suited to face the new reality in biological and ecological sciences (Marques, De Loof), in social sciences (Mesjasz, Spivak), and in economical thinking (Petrella, Mark, Rosicka).

The common denominator to all this material might be that too many concepts in mainstream sciences and in conventional thinking are outdated. According to these authors, the world in transition is badly in need of a paradigm shift.

Other contributors have focused their research on the emergence of the new (Robb, Peterson) or the novelties at the basis of the global transformations. Recent developments in information and communication technology (Petrella, Robb) are 'explosive' and transforming the socio-economical sphere, while humankind is at odds in handling an increasing number of uncertainties (Mesjasz, Spivak, Marques) and risk factors (Szántó). Starting from completely different grounds the impending ecological catastrophe can be avoided by wisdom or gnosis (Fennema and Szántó). For desequilibria of less grave extent 'indigenous knowledge' is revived (O'Reilly). Following from these diagnoses of humankind at high risks, the prescribed therapy schematically develops alongside the following three headlines: Well focused scientific research, socio-political action, and the call for a powerful new synthesis.

Some authors give indications for appropriate research programmes in their respective fields: in economics (Petrella, Mark, Rosicka), in philosophical conceptualisation (Aerts, Robb, Weyns), in the life sciences (De Loof), in necessary knowledge for survival (O'Reilly) and in scenario prospectives (Marques).

Other contributors very frankly transcend the cognitive to the ethical field, where they investigate the collective impact of individual hunger strikes (Landzelius), the empowerment or enhancing of individual autonomy and self-development (Szántó), or where they call for a new (eco)-ethics (Fu-Yang, Fennema).

Finally different contributors advocate a transgression of boundaries, be it between disciplines where they plead for interdisciplinarity (Marques, Weyns, Szántó, De Loof).

The inspiring cross-links between the presented reflections quite evidently impinge upon a number of pertinent debates on the social and eco-philosophical world scene. To name but four of them: The persistent tendency of the mainly 'private' globalisation of economy and communication remaining beyond public control [1]; The polemics about stability and diversity, and the emerging definition of ecosystem health and integrity [2,3]; The new naturalism or careful transition from the 'is' to the 'ought', from objective science to the evaluative and normative, in a rational way [4,5,6]; The issue of 'postnormal science' as a necessity to tackle high uncertainties, high risks and high stakes in a most rational way [7].

The number of associative thoughts raised by these combined reflections seems endless. Ultimately, this is a book about change, not unlike the *I Ching*—founded on ancient oriental knowledge of the archetypical cyclic changes in life—but more fundamentally the deep reshaping of reality on which the former variations are merely ripples. Definitely, the multicultural and interdisciplinary nature of the papers presented at the 'Einstein meets Magritte' Conference has produced a wealth of contrasting visions of high quality. Confrontations of this level are first hand opportunities to discover both the richness of cultural diversity and the possible overlapping consensus about a limited number of fundamental points. This is what we have tried to initiate in this short introduction. For guidance along this path and to this most challenging endeavour we propose the apron diagram developed by the Norvegian eco-philosopher Arne Naess:

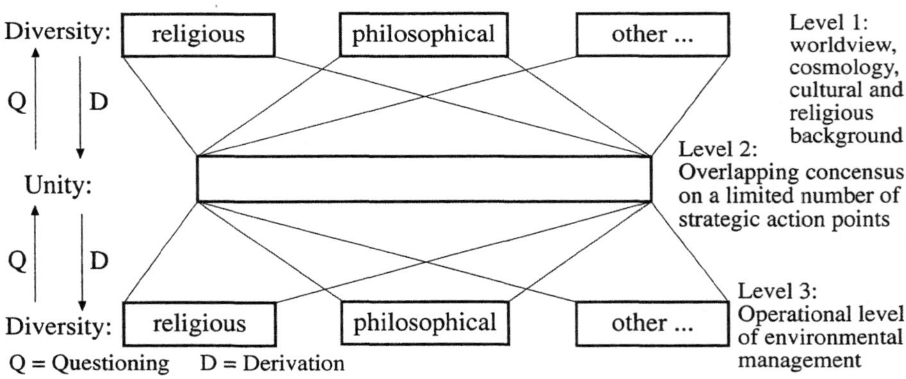

Diagram of overlapping consensus [8].

Inevitably this cultural and disciplinary richness has also a price, such as the outspoken heterogeneity of the system of cultural references of the individual contributors. Communicating with all writers disseminated all over the globe has been an almost toilsome task (even in the age of the world wide web). Therefore, we count upon your tolerance. According to us, it is worth it. The reader is hereby invited to confront the highly interesting standpoints and to judge by him/herself.

May this volume be an enriching and valuable contribution to those actual debates amidst others, so fundamental for the case of a highly qualitative co-development of humankind-in-nature.

Special thanks are due to many enthusiast participants and organizers of the "Einstein meets Magritte" conference. The authors especially are thanked for their valuable contributions, while all CLEA members and the 'text debuggers'—e-wizard Didier Durlinger and Sami Amira—are thanked for their continuous commitment.

Jan Broekaert
CLEA/VUB,
Krijgskundestraat 33,
B-1160 Brussels,
E-mail:
jbroekae@vub.ac.be

Willy Weyns
HUCO/VUB,
Laarbeeklaan 103,
B-1090 Brussels,
E-mail:
wweyns@meko.vub.ac.be

REFERENCES

[1] **OECD**, "Economic globalisation and the environment", june 1997.
[2] **Schrader-Frechette, K. et al.**, *Method in Ecology: Strategies for Conservation*, Cambridge, C.U.P., 1993.
[3] **Westra, L. et al.** (eds.), *Perspectives on Ecological Integrity*, Dordrecht, Kluwer Academic Publishers, 1995.
[4] **Apostel, L.**, *Wereldbeelden, ontologie en ethiek*, Antwerpen, Humanistisch Verbond, 1994.
[5] **Batens, D.**, "Rationality and ethical rationality", *Philosophica*, **22**, 1978, pp. 23-45.
[6] **Weyns, W.**, "The Cultural Filter", in: **Hens, L., Nath, B., Compton, P. & Devuyst, D.** (eds)., *Environmental Management in Practice*, (Vol 1), London, Routledge, 1998.
[7] **Funtowicz, S. et al.**, "Science for the Post Normal Age", in: **Westra, L., et al.** (eds.), op. cit.
[8] **Naess, A.**, "Ecosophy T", Appendix A, in: **Devall, B. et al.** (eds.), *Deep Ecology*, Salt Lake City, Peregrine Books, 1985.

1. **Social change and the individual condition.**
From a mainly economical stance **Riccardo Petrella** describes the evolving faces of the capitalist paradigm and the need for future redirection of global policies through new world organisations. De-emphasising competitivity and tools leads to rethinking the social global future.
In a systemic approach the barriers of cognitive evolution are investigated by **Borisz Szántó**. Not only the contradictory interrelatedness of the self-modifying cognitive progress of the individual—through education and emphasis on socio-technological possibilities—and the self-restoring but limited stability of the ecosphere, but also the linearity of an uncontrolled socio-technical environment sets boundaries to humankind's evolution.
A mainly regional Russian analysis of the social transition crisis during the post Cold-War period is made by **Dmitri Spivak**. 'Apathy', 'fear' and 'mysticism' are recognised as key concepts for transposing altered states of consciousness to socio-cultural groups.
A futurist speculation on emergent social freedom is drawn from insights in the ontology of quantum-mechanics and evolution by **Diederik Aerts**. Presented as a narrative account reminiscent to the Platonic dialogues, the individual's movement through social space and his emergent collaborations lead to a vision of an expanded social and interactive paradigm.
The subtle mechanism of the individual's ultimate contestation of social injustice is evocatively unfolded by **Kyra Marie Landzelius**. The mediatised image of the starving body as an icon of absolute ethics results from self-destructive hunger strike.
2. **Complex evolution and fundamental emergent transformations.**
In a thought experiment **Fenton F. Robb** explains how 'institutions' spontaneously emerge from social interaction as autopoietic and self-defining real-world entities. As supra-human 'institutes' define the human perception and understanding of the real, true, good and causally related, the need for research on this evolutionary change through 'Institutional Dynamics' seems necessary.
A major problem in transdisciplinary studies is the exchange of terminology and process mechanisms between disciplines, conveyed by analogies and metaphors. **Czesław Mesjasz** analyses the effects of the introduction of system analogies and metaphors in the study of contemporary world politics. Not only is the metaphor's normative value enhanced, but as social systems are self-referential and autopoietic they contribute to their proper development.

Dramatic changes in society, technology and environment enhance the emergence of pathogenic agents, increase the complexity of health profiles and affect the predictability of dominant health transitions. According to **Marilia Bernardes Marques** the concomitant intrinsic complexity of health policy assessment urges a transdisciplinary and technological prospective approach.

A careful analysis of the emergence of 'significant universals' in language and music is exposed by **Philip L. Peterson**. Endowed with explanatory capacity and biologically unreducible these cultural entities are considered to exist distinctly from nature.

A speculative symbolic formalism to implement the idea of communication in biology is advanced by **Arnold De Loof**. The analysis, based on non-linear dynamical systems, amounts to the claim that 'communication between hierarchical constituents' must be considered essential in the concept of 'life'.

3. **Ecological transformation and responsibility inquiries.**

The subjective construction of scientific truth and the need for lawful mediation between concomitant power and resistance is approached theoretically by **Serge Gutwirth**. The privileged role of science in environmental regulations now submitted to truth-pluralism and law mediation could—contrary to its previously alleged destructive role—be democratically guided.

By means of a SWOT-analysis **Willy Weyns** describes positive and negative potentialities of humankind's techno-ecological relations. The break down of successful micro-level strategies when transposed to the global scale, calls for new solutions and revision of the contemporary paradigm.

From a spiritual stance **Jan Fennema** builds up a reflection on the fundamental integration of humans in nature. Deepened ecological knowledge or Wisdom—more than Promethean arrogance—is needed to avoid further environmental deterioration.

Peng Fu-Yang presents a moralist analysis of the alienating influence of over-emphasised science and technology development. Support is given to policies for co-evolution of humankind and nature. Paradoxically this is proposed by means of scientific achievement and at risk of sacrifying the individual integrity.

4. **The economic-ecological and socio-technical equilibria.**

Prior to the industrially and commercially induced deterioration of the environment, 'Indigenous Knowledge' achieved a sustainable and harmonious relation to the ecosystem. **Francis D. O'Reilly** now advises 'necessary knowledge' to strategically counter the globalised economical pressure on Third World countries.

H Thomas R. Mark describes how the economic and ecological global system are both systems of hierarchically nested processes. The decoupling of both, causing depletion and deterioration of the environment, could be reversed through monetary institutions enforcing appropriate spatio-temporality on the global process.

The paradigm of sustainable development has been bypassed by the information society paradigm, ever relaunched by innovative technological achievement. **Riccardo Petrella** traces back the origins of the presently dominating paradigm to the capitalist industrial society paradigm. Applied differently, information and communication technology could facilitate the advent of a new global social contract.

In a historical survey of economy **Janina Rosicka** describes how the myth of fertility evolved into the opposed concepts of Scarcity and Plenty. The former became the axis of modern economy while the latter, reflected in the great social project, amounted to the concept of equality albeit on substantially low level.

Clearly all contributions are primarily concerned with the deep transformations of humankind and of the relationship between humans and nature. As a matter of fact, most contributions to the green book share what we are inclined to call a 'broad ecological vision' in their diagnostic of an increasing imbalance between humankind and its environment. Fortunately, as the reader going through this record will soon discover, this well documented factual statement is not their last word. The transition from the 'is' to the 'ought', from 'pure science' to ethics is generally made without hesitation. Almost all authors derive from their scientific description of the illness at least an indication of a remedy, each in their specific field of interest or discipline. No doubt, the multi-disciplinary background of the scientific contributors is one of the strong points and the originality of this truly international volume. The contributions presented here are from specialists in health research, biology, scientific ecology, philosophy, anthropology, sociology, physics, economy and information technology. Hence it is no surprise that the diagnoses of the ecological predicament and social transformations formulated by those distinguished contributors is extremely diverse, at least in first approach. Precisely these wide-ranging approaches allow creative associations between different contributions.

Some of the authors point at the human's belief of the myth of nature, and blame the 'Promethean arrogance' of the human species as a primal cause for the current state of affairs (Weyns, Fennema). Sometimes even intervention is proposed at the basic human level, putting first ecological integrity at the expense of individual integrity (Fu-Yang). Instrumentalism, defined as the precedence of the scientifico-technical realm over ethics is accused (Petrella, Mark, O'Reilly). Generally accepted concepts are not any longer suited to face the new reality in biological and ecological sciences (Marques, De Loof), in social sciences (Mesjasz, Spivak), and in economical thinking (Petrella, Mark, Rosicka).

The common denominator to all this material might be that too many concepts in mainstream sciences and in conventional thinking are outdated. According to these authors, the world in transition is badly in need of a paradigm shift.

Other contributors have focused their research on the emergence of the new (Robb, Peterson) or the novelties at the basis of the global transformations. Recent developments in information and communication technology (Petrella, Robb) are 'explosive' and transforming the socio-economical sphere, while humankind is at odds in handling an increasing number of uncertainties (Mesjasz, Spivak, Marques) and risk factors (Szántó). Starting from completely different grounds the impending ecological catastrophe can be avoided by wisdom or gnosis (Fennema and Szántó). For desequilibria of less grave extent 'indigenous knowledge' is revived (O'Reilly). Following from these diagnoses of humankind at high risks, the prescribed therapy schematically develops alongside the following three headlines: Well focused scientific research, socio-political action, and the call for a powerful new synthesis.

Some authors give indications for appropriate research programmes in their respective fields: in economics (Petrella, Mark, Rosicka), in philosophical conceptualisation (Aerts, Robb, Weyns), in the life sciences (De Loof), in necessary knowledge for survival (O'Reilly) and in scenario prospectives (Marques).

Other contributors very frankly transcend the cognitive to the ethical field, where they investigate the collective impact of individual hunger strikes (Landzelius), the empowerment or enhancing of individual autonomy and self-development (Szántó), or where they call for a new (eco)-ethics (Fu-Yang, Fennema).

Finally different contributors advocate a transgression of boundaries, be it between disciplines where they plead for interdisciplinarity (Marques, Weyns, Szántó, De Loof).

The inspiring cross-links between the presented reflections quite evidently impinge upon a number of pertinent debates on the social and eco-philosophical world scene. To name but four of them: The persistent tendency of the mainly 'private' globalisation of economy and communication remaining beyond public control [1]; The polemics about stability and diversity, and the emerging definition of ecosystem health and integrity [2,3]; The new naturalism or careful transition from the 'is' to the 'ought', from objective science to the evaluative and normative, in a rational way [4,5,6]; The issue of 'postnormal science' as a necessity to tackle high uncertainties, high risks and high stakes in a most rational way [7].

The number of associative thoughts raised by these combined reflections seems endless. Ultimately, this is a book about change, not unlike the *I Ching*—founded on ancient oriental knowledge of the archetypical cyclic changes in life—but more fundamentally the deep reshaping of reality on which the former variations are merely ripples. Definitely, the multicultural and interdisciplinary nature of the papers presented at the 'Einstein meets Magritte' Conference has produced a wealth of contrasting visions of high quality. Confrontations of this level are first hand opportunities to discover both the richness of cultural diversity and the possible overlapping consensus about a limited number of fundamental points. This is what we have tried to initiate in this short introduction. For guidance along this path and to this most challenging endeavour we propose the apron diagram developed by the Norvegian eco-philosopher Arne Naess:

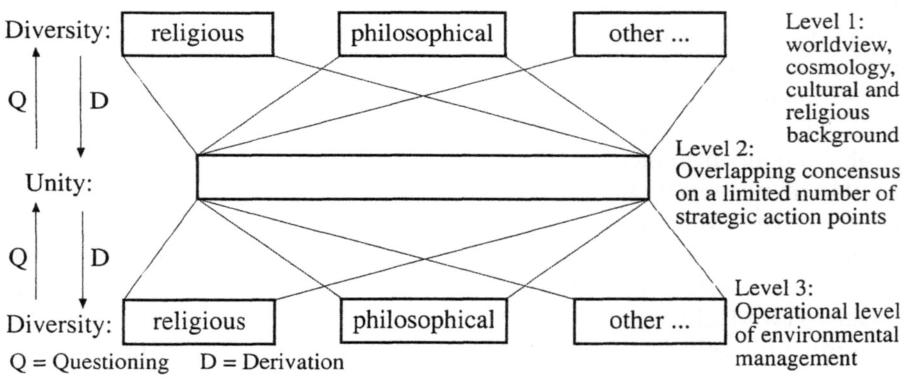

Diagram of overlapping consensus [8].

Inevitably this cultural and disciplinary richness has also a price, such as the outspoken heterogeneity of the system of cultural references of the individual contributors. Communicating with all writers disseminated all over the globe has been an almost toilsome task (even in the age of the world wide web). Therefore, we count upon your tolerance. According to us, it is worth it. The reader is hereby invited to confront the highly interesting standpoints and to judge by him/herself.

May this volume be an enriching and valuable contribution to those actual debates amidst others, so fundamental for the case of a highly qualitative co-development of humankind-in-nature.

Special thanks are due to many enthusiast participants and organizers of the "Einstein meets Magritte" conference. The authors especially are thanked for their valuable contributions, while all CLEA members and the 'text debuggers'—e-wizard Didier Durlinger and Sami Amira—are thanked for their continuous commitment.

Jan Broekaert
CLEA/VUB,
Krijgskundestraat 33,
B-1160 Brussels,
E-mail:
jbroekae@vub.ac.be

Willy Weyns
HUCO/VUB,
Laarbeeklaan 103,
B-1090 Brussels,
E-mail:
wweyns@meko.vub.ac.be

REFERENCES

[1] **OECD**, "Economic globalisation and the environment", june 1997.
[2] **Schrader-Frechette, K. et al.**, *Method in Ecology: Strategies for Conservation*, Cambridge, C.U.P., 1993.
[3] **Westra, L. et al.** (eds.), *Perspectives on Ecological Integrity*, Dordrecht, Kluwer Academic Publishers, 1995.
[4] **Apostel, L.**, *Wereldbeelden, ontologie en ethiek*, Antwerpen, Humanistisch Verbond, 1994.
[5] **Batens, D.**, "Rationality and ethical rationality", *Philosophica*, **22**, 1978, pp. 23-45.
[6] **Weyns, W.**, "The Cultural Filter", in: **Hens, L., Nath, B., Compton, P. & Devuyst, D.** (eds)., *Environmental Management in Practice*, (Vol 1), London, Routledge, 1998.
[7] **Funtowicz, S. et al.**, "Science for the Post Normal Age", in: **Westra, L., et al.** (eds.), op. cit.
[8] **Naess, A.**, "Ecosophy T", Appendix A, in: **Devall, B. et al.** (eds.), *Deep Ecology*, Salt Lake City, Peregrine Books, 1985.

RICCARDO PETRELLA

A CRITIQUE OF THE COMPETITIVENESS IMPERATIVE AND CURRENT GLOBALISATION

THE PURPOSE OF THE ECONOMY TODAY

The economy seems to have increasingly lost any sense of purpose. Gone is the goal of reconstruction that was the uppermost in the post-war period. Gone too is the goal of growth for its own sake, which has forfeited all credibility since the 1960s and is nowadays defended by nobody. It was replaced as of the 1970s and 1980s by the aim of 'qualitative growth'; but for hundreds and hundreds of millions of people it is a mere delusion, being regarded as the latest invention of the rich societies of the North in their own interests! The same goes for the goal of development. After these long years of crisis (seemingly in no hurry to go away), of large-scale unemployment (particularly among those under 25 and over 50 years of age), of growing poverty (over 80 million people affected in the OECD countries), of cities crammed and jammed almost to choking point, of drug addiction and crime (including State crime), the soul-searching goes on everywhere. Development of what? Economic development? But of whom and by whom? Social development? But where is it? For that matter, are economists and industrialists equipped to talk about this and can they in fact do so?

This loss of purpose seems deep-seated, for never has there been so much debate about ethics and economics, ethics and business or the enterprise and ethics as there is today. And yet ...

In reality, economics has not lost its meaning. It is prompted, guided, directed by a purpose that now takes precedence over all others. It is that of competitiveness (of the short-term, commercial brand in particular) that has become the only true objective—sold, propagated and defended—of the dominant economy of the 'Norths' of the planet.

While, in economics, competitiveness is just one facet of the behavior of economic actors, in the context of supposedly competing markets it is no longer a means, it has become a prime objective not just of enterprises but also of the State and of society as a whole. When it comes to reforming the education system, for example, you will invariably be told that the chief aim of schooling, vocational training and higher education is to prepare high-performance human capital in order to make the country's economy more competitive in relation to the foreign rival.

© 1999 *Kluwer Academic Publishers. Printed in The Netherlands.*

The 'gospel of competitiveness', like all ideologies, boils down to a few simple ideas. We are engaged willy nilly—so the industrialists, economists, political leaders and academics tell us—in a ruthless technological, industrial and economic war encompassing the entire planet. The aim is to survive, and survival hinges on being competitive. Otherwise there is no short- and long-term salvation, no growth, no economic and social welfare. No autonomy, no political independence. The chief role of State, local authorities and trade unions is to provide the most suitable environment for enterprises to be, become, or stay competitive in the world economic war.

How is it that ideology or cult of competitiveness has forsaken the narrow bounds of the industrial and financial world, of the blind theorists of the market economy and the invisible hand, and of the fanatical thurifers of vulgar economic Darwinism to permeate world opinion right into the places most affected by destitution and social exclusion?

Of the many causes of the phenomenon, two are noteworthy here: prime emphasis on the tool, following the triumphant 'technologization' of society in the last 40 years; and the universalization of the economy coupled with the 'crisis' of the State as vehicle, promoter and guarantor of the public and general interest.

PRIME EMPHASIS ON THE TOOL

Such emphasis comes as no surprise. Nor is it entirely groundless. There can in fact be no denying, other than through naivety or blindness, that the scientific and technological advantages of recent decades have marked our time and are much to the fore in shaping the future of societies and of the human condition. Whether we think of production engineering (digitally controlled machine tools, robots, computer-assisted design and manufacturing, expert systems, etc.), biological engineering and medical technologies (scanner, magnetic nuclear resonance, lasers, optical fibres, image processing, etc.) or space (space shuttles, satellites, remote sensing, etc.), no human activity breaks free of science and technology.

Even our language testifies to this. We speak of such things as computer illiteracy, communication highways, self-reproducing robots, computer sex, test-tube babies, intelligent houses and virtual imagery. The celebration of science is matched only by 'technological eagerness'. Local officials, for example, dream of nothing sweeter than establishing science cities and technology parks, marking out technological routes and inaugurating teleports. So everything goes to build up incantatory discourse and spur investment in huge prestige technological projects announcing to some and promising to others a better world to come by dint of ever more powerful, expensive and high performance tools.

It is the context that the developed countries have well and truly opted for the culture of objects characterized by visions of the future and short- and long-term action strategies seeking to *promote tools rather than persons* and means rather than ends. The development of tools is even looked upon as the pre-requisite for human development, and the means stand in for the goal. Even though the emphasis on quantity (more cars, refrigerators, computers, satellites, planes, medicines, and so on) has in recent years been eased, it is still the case that the priority goal remains, legitimately, the growth of markets through the growth of products, tools and machine systems and universes supposed to increase the satisfaction of individual and collective needs.

The field of telecommunications is an enlightening example among so many others. With the growing combinatory use of computers, optic fibers and satellites, the postal, telegraph and telephone services and the giants of the telecommunications industry have embarked on the adventure of creating in the next 5 to 10 years a major national, European and global system of broad-band communication 'highways' known in technical jargon as ISDN (Integrated Services Digital Network)—others speak of 'Integrated Broadcast Communication Network (IBCN)' – while nobody seems to know what types of new services such a system will be able to provide in addition to what is already offered by Internet and what public needs it will be meeting. The argument of the promoters is a classic one: the predominance of the *object* over the nature of the service has to do with the temporary underdevelopment of the service. There can be no telling, it is said, what the tool will be used for until the objects and their auxiliary equipment have been made extensively available and the range of users filled out. Hence the precept "Get out the objects and the rest will take care of itself!"

Innovation and development in European societies are thus (in many respects, once more) defined in terms of 'technological' tools and systems'and no longer in terms of social good and services. The 'affirmed' rationality of the system and the reasoning as to the relevance and efficacy of the 'higher-performance' tools legitimates the processes of earmarking available material and immaterial resources.

The culture of innovation is chiefly a culture of managing technology. As virtually all management and business schools constantly teach, innovation is thought out and experienced as a corpus of technical, economic and financial data and factors. At the very most, some receptiveness to management sciences is noted. Human and social sciences take a back seat.

The human person is reduced to a '*resource*' when part of the production process and to a '*consumer*' when part of the utilization process! The rest is just marginally and must remain so, noise (which can be very costly) or even luxury (which, given sufficient wealth, can be indulged in).

That being so, when engineers, economists, bureaucrats, or even intellectuals of 'hard scientist' or 'soft thinker' brand make reference to an innovatory process or product, the qualifiers they use relate in the vast majority of cases to the production and management of technical tools and systems. The keywords used to describe an innovatory society are:

- productivity	- competitiveness
- efficiency	- profitability
- flexibility	- adaptability
- measurability	- optimization

To the extent then that the 'technological' media and tools become part and parcel of human activities (can we nowadays conceive of a factory without computers?), any human activity is rationalized, re-expressed as inputs and outputs, evaluated in terms of solely monetary cost and benefit or goods, and exchanged like any raw material or industrial product. The human body itself is increasingly subjected to this kind of strength.

RESTRICTED GLOBALIZATION OF THE ECONOMY: THOUGHT OUT AND MANAGED IN THE EXCLUSIVE INTERESTS OF 800 MILLION SOLVENT CUSTOMERS

The foregoing is situated in and interacts with another major phenomenon of contemporary history, namely the nascent globalization of the economy and of society.

Globalization of the economy is a new phenomenon in relation to the old processes of internationalization and multinationalization. It covers not only exchanges of goods (raw materials or finished articles) and services (as in internationalization) and capital transfers associated with extension of the activities of economic units (enterprises in fact) in other countries of the world through the establishment of direct subsidiaries or by means of acquisitions (as in multinationalization). Globalization concerns all stages of design, development, production, distribution and consumption of goods and services. Credit cards, the motor industry, 'fast food' catering, air traffic control and seat reservations are typical examples of globalized goods and services. It is to be noted that globalization does not necessarily mean standardized products and services for uniform world markets: the same car, the same chocolate. Quite the opposite.

What is still more significant is that globalization differs from internationalization and multinationalization in two other respects. The first is the fact that globalization results in the end of national industrial capitalism marked by the gradual disappearance of the principles, rules and modes of organization and enhancement of natural and immaterial resources of the 'national' system: national State, national economy, national currency, national bank, national postal and telecommunications services, national markets, and so on. Globalization spells the end of the national as a strategic dimension and an exit and entry point for economic units. The second difference is that globalization of the economy results, for the time being, in a 'system' governed chiefly by a private actor, the enterprise as a world network, which was not the case in the internationalization and multinationalization period. Globalization escapes the national public authorities, though without coming under the control of world public authorities since none exist. It 'simply' comes under the control of worldwide networks of firms and conglomerates of giant firms, the only structures that have so far been capable of achieving globalization.

Globalization of the present economy is nevertheless something narrowed down since it does not involve visions, strategies and action by and in the interests of the world population. The world map as seen, mentally and culturally, by the people and leaders of the rich countries of the North, particularly by the Japanese, Western Europeans and Americans, is a distorted map. Their impression is that the world is getting ever smaller and more unified around their poles. They think that the world that counts, in present and future economic, political and cultural development terms, is made up of the triad formed by the United States, Japan and Western Europe. That being so, the big issue of the next decade or so is which of these three entities will by the year 2010 be the world technological and economical leader. All the rest is mere agitation on the sidelines. Rather than veritable globalization, we should for the moment be talking about the 'triadization' of the economy.

The implications and consequences of the combination of overemphasis on tools and a narrowed-down (triadic) globalization of the economy under the governance of a private main actor (the enterprise) are considerable and many. Three of them mark our era: (1) we live in the North with an attitude of survival dictated by a supposedly inevitable context of worldwide economic war; (2) we are witnessing a new alliance between enterprise and State resulting in, among other things, the assumption by the former of the task of determining the substance and limits of the general public interest; and (3) we are being subjected to a dismantling of the social contract that underlies the economic and political development of Western societies, and hence to a universalization of social exclusion.

THE ATTITUDE OF SURVIVAL IN A DECLARED CONTEXT OF INEVITABLE ECONOMIC WAR

The first series of implications and consequences is cultural. They concern the domain of perceptions, aptitudes and beliefs, the realm of the mood.

With the extent (and acceleration) of the transformations linked to the scientific and technological changes to which industries, organizations and countries are 'subjected' in an increasingly vast, worldwide and hence harder-to-manage context, the feeling of being up against a process of permanent destabilization in an environment of growing uncertainty and hostility is becoming widespread.

That being so, we have the impression of being caught up in an attitude of survival for survival's sake (remain in the market; keep our job and what goes with it; prevent the foreign competitor or just the immigrant from taking our place). 'The other' immediately becomes suspect as a potential source of destabilization, hence a threat to our survival, even in the deepest recesses of our identity.

Having the most efficient and cheapest tool that meets the needs of the richest, most solvent markets becomes, in this context, the best means of ensuring (micro-) survival. Salvation lies in the tool! Innovation on and by the tool and competitiveness in and by the price and quality (defined in technical and financial terms) of the tools emerge as the only effective and apparently legitimate regulators of the war for survival.

The prevailing thinking on innovation and the obsessive cult of competitiveness have had a big hand in developing and spreading this 'survival psychosis' and 'war culture'. The values emphasized by the theorists and practitioners of innovation and globalization through competitiveness are values centering on power (financial in particular), force (idea of conquest and mastery), struggle (to which the inevitable instances of co-operation between individuals, organizations and countries must also be subjected), aggressiveness and cynicism (*mors tua vita mea*). Admittedly, one could say that there is apparently nothing new on the scene of human history and of societies. Yet there is something new in that, for the first time, such an attitude has assumed a worldwide, global dimension—that of the totality of human history, in a context of real potentiality, on the one hand, and of the total self-destruction of human society (as a result of nuclear energy) and of the general manipulation of the living (with the new advances in genetic engineering), on the other.

THE NEW ALLIANCE BETWEEN ENTERPRISE AND STATE: A 'REVOLUTION' OF GOVERNANCE

Being regarded as the principal force in technological, industrial and economic innovation on which the future of a city, a region or a country depends, and as the only organization capable of 'managing' the globalization of the economy, the enterprise is becoming the force which makes culture, i.e., which shapes the value systems of a society and an era and fixes the rules of the game and the way in which the world economy is to be steered and managed with the backing of the State and of all the public authorities (local, national and international). There are compelling interests on either side in favor of the alliance.

The enterprises need the direct and indirect support of 'their' own State (or even of 'their' multiple States when they are present and play a significant role in the economies of several countries) to succeed in and 'control' their globalization. Beyond the official approach of 'less State, more market', they ask the public authorities to protect them at home (e.g., through the award of public contracts) and to ensure that they have a place in the world vanguard, particularly in high-risk or the most promising sectors (e.g., through the direct or indirect soft financing of their basic R&D programs). In return, the enterprises assure the State that they will become or remain competitive, thereby contributing—they say—to the creation of jobs and to the country's technological autonomy.

States, for their part, have an objective, direct and immediate interest in the alliance between enterprises and in the support they give to the enterprise of the country. Given the 'national' independence and economic welfare increasingly depend on innovations in advanced basic technologies and the marketing of the products and services deriving from them, the State is to an ever greater extent 'tied in' with the technological innovation and 'control' of the markets that is ensured by the enterprise. Since its political and social legitimacy—depending directly on its ability to ensure the nation's socio-economic development—is at stake, is has every interest in supporting its own 'national' enterprises.

From one OECD country to another, in different forms, the same logic thus operates, with available resources being mobilized for the sake of the short and medium term commercial success of 'national' enterprises, particularly the strongest, the 'winners' in world markets. What we have is a massive transfer of public resources in favor of private enterprises, mainly multinational, to enable them to stay competitive in the most profitable solvent markets.

The enterprise is acquiring anew historical legitimacy inasmuch as it has been vested by the State with the function of defending and promoting the well-being of the 'local' society, it claims an additional legitimacy by presenting itself as the only organization capable of guaranteeing, at the world level, the best management of available material and non-material resources. In so doing, the enterprise privatizes and internationalizes for its own purposes the social role of the State. In addition, it does so repeatedly, in each of the countries where it can lay claim to a predominant place in the local landscape. What is more, and failing a World State, it also privatizes the function of organizing the world economy.

While the enterprise is thus assuming the prerogatives of the State, the State is increasingly behaving like an enterprise in backing the globalization when victorious, of its 'national champions'. This globalization increasingly places States in competition with one another. It is no mere coincidence that we now speak of 'Japan, Inc.' or 'Enterprise France' to signify that an entire country (and the corresponding social and political formation that makes the State) is reduced, at the world level, to conducting itself chiefly as an economic agent. Hence a fundamental inversion of roles, where it is those in particular sectors—and pursuing particular interests—who drive and direct the action of those who are discharging general functions and upholding the public interest!

This new configuration tolls the knell of competition and the famous 'free market forces'. Within each country first, particularly in the key sectors (telecommunications, robotics, aeronautics, chemistry, computer technology), the 'national champions' are relieved of any true competition; at the international level, we can no longer regard firms linked by a series of alliances or agreements, from the input stage of research and development right down the line to marketing, as genuinely competing. What 'competition' remains becomes intra-enterprise competition, as long practiced by many industrial groups in consumer goods: foodstuffs, washing powders, cosmetics, cars, etc.

We are thus seeing a gradual and growing dissociation between globalized economic power and national political power, organized at an inadequate strategic level in relation to the global space. Technology serves to amplify this dissociation and is less and less subjected to 'national' democratic oversight. The person in the street may feel increasingly powerless in the face of world phenomena and events, but the 'national' ruling classes are just as much at a loss. Everyone is finally aware of and 'accepts' the fact that the major enterprises and their networks have become the institutions that govern the world's technology and economy, as kinds of world States but without the responsibilities of such statehood. The situation is comparable, *mutatis mutandis*, to that of the thriving European economy of the fourteenth and fifteenth centuries, hinging on the cities and the alliances they built up among themselves. Today's multinational and global enterprises reflect many a feature of the German Hanse and take us into a 'Hanseatic' phase of the organizational mode of the world economy and global society.

Will this globalization allow the emergence of a transnational civil society, namely one boasting relatively shared principles and value systems and displaying forms of expression and organization of collective living that ensure new modes of civil regulation worldwide? It in any case urgently demands new forms of political regulation. For if brands of representative democracy are to retain an ounce of authenticity, the control of technology and the economy cannot be left in the hands of people just serving special interests.

Besides, reality speaks for itself since governance by the globalized enterprise is undermining the social contract the world over and accentuating the phenomena of regression and social exclusion.

IMPAIRMENT OF THE SOCIAL CONTRACT AND INCREASE OF SOCIAL EXCLUSION

The living standards achieved by Western societies (European and North American) rest on a 'social contract', a compromise built up over more than a century of struggle but also of a quest for consensus and solidarity between social classes, organized groups and regions within individual States, and between States. The contract in question consists of rules, both written and unwritten, practices, values and institutions governing four major areas:

- Right to work, established as a founding principle of the State by the post-war Constitutions of many Western countries, a case in point being Italy.
- Redistribution of wealth ensuring, through a system of social assistance, general living or subsistence resources for citizens occasionally or permanently unable to fend for themselves.
- Protection against social risks in order to ensure all workers and their families against such occupational hazards as sickness, accident and death. In many countries, this system of insurance or social security has been extended to the entire population and all of life's risk.
- Promotion of equal opportunity, with the fostering of conditions enabling all citizens to express their capabilities and follow their

inclinations irrespective of social origin or physiological status.

Yet the developments of the last 20 years—the crisis of the 1970s, the aging of population, the neo-liberal wave of the 1980s and the heightened bid for technological and economic leadership involving Japan, Western Europe and the United States—have called into question the terms and validity of the social contract. What future then, we may ask ourselves, can there be for it in the globalization of the economy?

The globalization of technology can admittedly do much in favor of the social contract, for example, it could:

- facilitate the transfer and use of safer (reduced accident risks) and more ergonomic (better working conditions) technologies in production work;
- permit the spread of practices under which workers take part in managing innovation (as in recent years with the Japanese-style quality circles);
- speed up modernization of the productive apparatus of poor or less developed countries and create new production opportunities there, particularly in the fields of energy, agriculture, biotechnologies and transport;
- encourage the introduction in labor legislation (statutes, conditions, pay) of standards and rules comparable to those in force in the European Union countries, thereby reducing the now very considerable scope for 'social dumping';
- stimulate the development of international pay bargaining (within individual branches or world networks of enterprises);
- contribute to a redistribution of wealth-producing capacity around the world and so combat poverty in African, Latin American and Asian countries, as well as in Europe and North America;
- foster the framing and application of a world employment policy, as outlined by Mr Louis Emmerij, former President of the Development Center of the Organization for Economic Development and Co-operation (OECD).

However, the drawbacks of globalization are just as considerable. Competitive globalization is used as an argument to:

- halt and/or cut back on legislative measures already taken in biotechnology for reasons of safety, quality and ethics, and prevent the adoption of further legislation;
- accelerate 'technological laissez-faire' for standards and the opening up of markets in the telecommunications sector;
- impair machinery for discussion and bargaining about technological change in the enterprise, such procedures being regarded as a luxury

in a so-called period of 'world technological war'.

In this respect, there can be no doubt that the globalization of competition by the advanced technologies will continue to be one of the factors of the ongoing crisis of 'claim-pushing' trade unionism in developed countries, and also one of the curbs on the advent of active and strong union activity in the developing countries. So further inroads on job security are likely in both developed and other countries.

To counter this risk, an argument is often advanced by the ruling classes of the developed countries that seeks to oppose employment to joblessness. In a situation of full employment, it is contended, the legitimate social objective is a guarantee of equitable social cover for all types of workers; but, with the worsening employment situation the world over, cutting unemployment becomes the prime public objective. The quest to make the national economy more competitive—by constituting 'national champions'—is then to be seen as one of the keys to job creation.

To ensure this competitive edge, some concessions have therefore to be made regarding worker protection so that enterprises facing cut-throat competition are no longer penalized by national 'rigidities' in the social or economic sphere. Hence the proliferation of insecure employment arrangements—such as fixed-term contracts—and the various flexitime schemes, together with the reconsideration of such gains as, in some countries, the fixing of a maximum wage.

The argument boils down to this: going back on the social advantages of employees is legitimate since it helps improve the overall employment situation by making the country's enterprise more competitive. As it happens, most national experience shows this argument to be simplistic and misleading. The neo-liberal politics applied for a full dozen years by Mrs. Thatcher in the United Kingdom were not matched by any conclusive economic performances, since inflation and the number of jobless remained above their 1979 levels. Furthermore, the United States, where public social protection is practically non-existent, has entered a period of 'structural' recession.

The promotion of stable unemployment is liable to become a vacuous formula in Latin America, Africa and Asia since, in the next ten years, those regions would need to create more than a billion jobs just to stabilize unemployment at its present-day level of 800 million. In these circumstances, a deterioration in working conditions is to be feared in the developed countries, while the situation is likely to worsen not only in the developing countries but also in the so-called newly industrialized countries of South East Asia, as a consequence of the recent crisis.

Increased workloads and tempos to raise productivity will result in greater stress and markedly less respect for the individual in the enterprise. The globalization of all phases of production will, in addition, lead to denationalization of social security systems through growing individualization of income, benefits and their funding. Such individualization, already under way in many developed countries (United States, United Kingdom, Germany, Belgium, Italy, France), will benefit high-income groups and the employees of multinationals and world networks of enterprises. On the other hand, it will result in a relative lowering of the level and quality of benefits to those with low incomes and employees of local enterprises. In this respect, women and migrant workers will, once more, be the most vulnerable social groups.

The globalization of production, capital and services in the context of out-and-out competition for world leadership between the three rich 'regions' (Western Europe, Japan and the United States) will also lead, in the next 10 to 15 years, to a strengthening of the most developed zones of those regions. This is enough to undo all the good of the regional policies applied between 1950 and 1980, and that the economic crisis of the 1970s had undermined as it was.

We may therefore be allowed some misgivings about the present significance and scope of the 'social contract', and a good dose of skepticism as to the capacity of competition-based globalization of the economy to curb or halt the rise of poverty in the developed countries in the next two decades.

FOR ECONOMIC DISARMAMENT: DE-EMPHASIZING COMPETITIVENESS

The economy must be disarmed. Entrepreneurs must be urged to stop dressing up as generals and colonels; economists to forsake their military strategists' garb or their scout caps; business and management schools to take a little well-earned rest after so many battle plans so arduously drawn up, discussed and sold; national public authorities to abandon their role as standard-bearers or panderers to reactionary forces; the highly reputable and influential economic and financial press, Anglo-American in particular (e.g., the *Financial Times*, *The Economist*, the *Wall Street Journal*, *Fortune*), to end their years of blowing the same hunting horns and putting out the same martial music summoning us to do battle over skies, oceans, cables, chips and investment funds.

Apart from that, there are quite a few important initiatives that can and must be taken to initiate and speed up economic disarmament, starting with the countries of the Triad.

One of the decisions that the Group of Seven most industrialized countries of the world should take is the formation of a 'Committee for World

Economic Disarmament', with an invitation to each member country to set up a National Committee for Economic Disarmament (to replace the numerous committees or national councils on competitiveness established in the 1980s and the 1990s).

It is a great pity that the world leaders have missed the opportunity to take advantage of the celebration in 1995 of the fiftieth anniversary of founding the United Nations to launch two initiatives: (1) organize, on a multilateral and worldwide basis, the first economic arms limitation talks; and (2) secure the approval, by the United States Security Council, of a Memorandum on Economic Disarmament for World Economic Security. These initiatives must be promoted.

In conjunction with these initiatives, a campaign should be run to de-emphasize the virtues of competition and of the many economic indicators that now compose the ABC of the political economy culture (such as GNP growth, labor productivity, enterprise competitiveness or the competitiveness of a national economy or a region) and feed the culture of economic welfare. In their place, a new set of human and social development indicators should be spotlighted, as called for in studies by UNDP, the World Bank and many non-governmental organizations. Academics and scientists have an important part to play in this respect.

Another initiative is needed. As in the case of the EUREKA program, which awards a 'Eureka' label for inter-enterprise co-operation projects meeting the criteria laid down by the EUREKA Member States, the United Nations should, in co-operation with a reshaped World Bank and International Monetary Fund, award a 'World Priority' label (with all the financial and prestige benefits involved) to activities and projects of enterprises in line with the objectives of co-operation, co-development and solidarity. Also eligible for the label would be comparable initiatives by chambers of commerce and industry and by cities, together with joint courses by universities of different countries of the world, as exemplified by the SOCRATES—formerly ERASMUS—program. Universities, for their part, could launch a joint world campaign by including in their curricula, 15 or so courses on economic disarmament. Foundations could back such a program.

Furthermore, these measures must fit into a broader framework requiring greater commitment, that of a world social contract.

SEEKING A WORLD SOCIAL CONTRACT

The reference and starting point of the world social contract are the needs and expectations of eight billion human beings who, in 22 years' time, the space of a generation, will be living on this earth. Given present trends, there will be some three billion poor (with a daily income of less than 1 US dollar), over 2.5 billion homeless and two billion without mains water, and so on. Africa could vanish from the world economic map.

In such circumstances, satisfaction of the basic needs and expectations of eight billion people is the only true and compelling objective for economic and social development in the next 22 years. It is therefore time to recognize the right of every individual to enjoy access to an income making it possible to live in society and providing for housing, food, medical care, education and clothing.

The world social contract will need to be based on two categories of action: (1) the adoption of new worldwide economic rules of the game, in an appropriate legal framework; and (2) the preparation of a series of instruments and measures ensuring effective implementation.

As regards the first category, the increasingly global economy of this close of the century cannot possibly form the effective basis of worldwide development without also a globalized legislative and political framework. Since a world State cannot be envisaged in the foreseeable future, work must go ahead on determining the basic principles, major thrust and substance of a veritable world contract at least covering:

- *The identification of common goods and services of collective property* forming the heritage of the whole of humanity, the management and betterment of which should obey common rules of use and access the world over. This includes natural resources (such as water...), certain basic technical infrastructures (telecommunications...), and scientific knowledge.
- *The creation of a world socio-economic citizenship* to accompany the migratory flows that will be intensifying, if only on account of the population explosion. Concretely, this means gradually promoting a world policy of labor and population movements. Following the example of schemes that have succeeded elsewhere, like SOCRATES or the Fullbright scholarships (enabling researchers and academics to complete their training or conduct research in the United States).
- *The design and launching of a number of major solidarity projects* aimed at preventing a complete hiatus between the economies and societies of the 'North' and 'South' of the earth. The dominant economies of the twentieth century were built up as a result, *inter alia*, of vast collective development programs involving free and compulsory schooling, general availability of health care, low-income housing and major road-building programs. Without all that, the poor cities and regions of the North would have become further weighed down by their backwardness or handicaps. What is true at

national level applies equally to the world as a whole. It is timely to implement a Global Water Contract aiming at recognizing water as a common patrimonial good of mankind and, therefore, access to water as a basic sacro-economic right of each citizen and community (see also infra where reference is made to a contract for home, water and good).

- *The development of new forms of economic, scientific and technological, and socio-institutional engineering co-operation projects*, conducted and managed by specific task forces for each co-operation project. The aim would be to apply worldwide various practices in force in many countries (Japan, Germany, United Kingdom, United States, Switzerland, etc.) for funding large-scale research and development programs, the establishment of economic growth points and the implementation of multinational projects such as Airbus and the Channel Tunnel.

As to the second category or proposals, faster ways and means must be found of managing the world economy, for that is still a long way off indeed. It is more the case that the present globalization condemns us to out-and-out wastage of resources to gain the competitive edge as the vying for supremacy goes on between the United States, Japan and the various countries of Western Europe (the latter still incidentally in 'extended order'). To achieve all this, the following action is required:

- restoration of national, continental and *global checks on capital flows*. Such movements have become entirely free worldwide, which has left the door open to unbridled speculation, suicidal instability in money markets, proliferation of a jungle of directionless financial services (the derivative markets), volatility of returns, and so on;
- the *imposition of taxes on capital gains*, particularly of the speculative kind. Never have the banks and insurance companies of the rich and poor countries shown such high profits as in 1993 and 1994 despite the greater numerical incidence of poverty in the world;
- the *abolition of tax havens*: there exist 37 'Tax havens' throughout the world;
- the immediate application of a *compensatory social levy on delocalized economic activities* in order to constitute a world fund for economic and social development in accordance with the new social context;
- the *introduction of 'eco-taxes'* enabling the focus of taxation, at present centered on human work, to be shifted to other factors of production such as energy, capital investment and polluting raw materials, which would provide an incentive to make energy, plant

and capital more productive and so de-emphasize increased labor productivity; the only ecological concern in the future planning for 22 years;

- the *establishment* within the United Nations *of a World Anti-Trust Authority* to be responsible for preventing the formation of large world financial and industrial oligopolies concentrating in their hands enormous powers to influence the world economy, lying beyond any system of democratic oversight and nullifying any economic action measures on the part of national and continental authorities;
- the *constitution of a World Economic Tribunal* to secure observance of ethical principles in world economic affairs (for which purposes the Tribunal should produce an annual report on 'Ethics and business in the world') and to act as an audit office regarding the use of world public financing for cooperation and development.

All these proposals require new institutions and new machinery subtended by principles no longer those of WTO, the IMF and the World Bank.

What is needed for this purpose is to design and establish a governance of the world economy by going on from today system made of WTO + IMF + WB + ILO + G7 to the WSDO, that is, the establishment of a World Social Development Organization (WSDO). This should be the outcome of a reorganization or integration of the WTO, the IMF, the World Bank and the ILO, also incorporating UNCTAD and the main forces behind it, overlooked by the rich countries in the last 25 years.

In its initial years (start-up phase), the WSDO would have three main tasks:

- *establish*, on the basis of a negotiated partnership with specific world networks of multinational financial and industrial enterprises willing to serve as 'citizen enterprises', *a program to rebuild the basic wealth of the world population* in houses, mains water, power supply systems, food, health and transport infrastructures, information and communication, and education by taking advantage of the necessary reconversion of military industries permitted by the end of the 'Cold War' and the East-West cleavage of the world;
- organize its own functioning on the basis of *a large variety of decentralized and regionalized mechanisms*, particularly in connection with the management of investment funds and development projects, with the establishment of inter-regional co-development networks between cities and regions of the poor countries;
- *promote the worldwide redistribution of resources* for the benefit of

the most destitute populations, notably through the promotion of funding of world youth employment measures based on small-scale schemes capable of involving the young of the place concerned.

In this setting, it will be for the WSDO to back and bolster a multitude of local development initiatives, both existing and potential, by innovating in regard to forms of funding and remuneration. For example, a contribution could be made to improve the living conditions of children under 12 in the poor countries by offering, in exchange for discontinuance of their work, to cover the cost of their education with funds supplied by the rich countries via the WSDO.

Access to work and assimilation of the technological hinge on the capacity of each country first to meet its basic needs by using its local resources and, in the process, drawing on the skills and technologies of the more developed world. For this reason, the world social contract should result mainly in priority for: (1) a contract for home, water and food; and (2) the rebuilding of towns and cities.

The contract for home, water and food. There is no society or human and social development without human beings living in a place somewhere, in a community. Yet more than 1.7 billion people at present have no home. One of the priority tasks for today's first planetary generation is therefore to live on earth, with the objective of providing homes for over a billion people by the year 2020.

It is in cities, in particular, that economic security through work and the need to dwell somewhere—key features of the world strategy for social development—find one of their crucial places of response and realization, yet cities are in a state of crisis and dying. It is predicted that in 2025, upwards of one billion Asians will be living in 50 cities of over 20 million inhabitants: an impossible situation. Thought thus turns to a new 'ecological' city, to the 'sustainable' city, to the 'agora city'.

It is proposed that, on the initiative of the associative movements of 50 cities in the world, a world inter-city network should be set up for the purpose of framing and securing the signature of a Federal Inter-City Covenant for the 'Urban Works of the Twenty-First Century'. The Covenant should consist in establishing the necessary conceptual, financial, institutional and managerial means for carrying out urban reconstruction programs in 50 to 100 cities of the world (preferably with populations running well into seven figures). Each 'Urban Work of the Twenty-First Century' should be based on employment of the local population of the districts to be rebuilt. The enterprises involved in the works as financial and industrial partners will receive the 'label' of 'world interest enterprises', giving them specific economic, tax and social benefits.

The twenty-first century will bring us into the third millennium AD, the millennium of the start of planetary history. We have to think of it with a great deal of hope and a great deal of peace in our minds. So conquerors need not apply; just stay put! Hegemony is definitely out. But builders of new boroughs and agoras and of new cities of the planet, please step right in.

Riccardo Petrella
President of 'the Group of Lisbon'
Commission of the European Community
Brussels, Belgium

BORISZ SZÁNTÓ

BARRIERS OF HUMAN EVOLUTION

INTRODUCTION

([1]) Theoretically the barriers of human evolution are to be derived from the phenomenon and rules of socio-technical functioning, its interactions with the human made functional environment.

The analysis propounds that most of the *conflicts and crises*, that humanity may expect in the future, are of *anthropogenic* character.

The irreconcilable contradictions, like the one between the self-modifying nature of human being and the self-restoration capability of nature, might become barriers on the road of *cognitive evolution.*

THE COGNITIVE EVOLUTION AND ITS PRINCIPLES

The stages of cognitive evolution in the individual can be differentiated according to one's age, heredity, motivation, experience and cultural environment, but it is the personal qualitative mental capacity and not the quantity of knowledge acquired that really counts [11]. The evolution of the cognitive capabilities of an individual is in correlation with *technological and cultural* development of his society. The society influences the mental development of a person, forces him to change by adaptation to mutable social circumstances and conditions. On the other hand, the individual's outstanding cognitive result may propel onward the development of *the whole of mankind.*

Piaget points to an inevitable parallel between biological evolution and cognitive development in the life of an individual [4] [5]. The development comes forth through periodic assimilation of new modes of function performance by whatever structure it has, followed by its accommodation and equilibrium [12]. The function we are speaking about is the problem-setting and problem-solving activity of the intellect.

According to the theory of neurophysiological system functioning, developed by the Russian neurophysiologist Piotr Anokhin [2], living creatures thank their existence to interdependent transient systems, being mobilized from various organs, cells, tissues, etc., of their organism, as

[1] Published with permission of the journal Systems Research and Behavioral Science, Lincoln, UK. Reference to Syst. Res. Vol. 15, (1998)

© All Rights Reserved.

well as some components from outside the organism, unified into an entity by an effort to pursue the projected result to satisfy this or that need of the organism. The purpose is being formulated within the system itself and manifested by a model of the future result anticipated and modelled by the central nervous system ('preceding reflection of the reality'). With the result achieved, the system disintegrates, ceases to exist, leaving the memory of its functioning in the central nervous system and in former components, modified in the process of functioning. The system gives up its activity to make space for the next gradual mobilization of its own or other systems, combined totally or partly from the very same components. Each integrating functional system interacts with its environment as an entity.

Evolution is the *criterion* of such a functional system, not its outcome or property. The functional system, its natural nonlinear totality, its feedback, oscillation, cyclic, result oriented, deterministic (*teleonomic* rather than teleologic) mode of operation is the manifestation of evolution itself.

Right from the beginning of its functioning, the structure of a *socio-technical functional system* even in its cognitive form can be given by the symbolic expression

$$[\textit{Human Being} + \textit{his Means}]$$

The active amplifying core and passive linear parts correspond to the structure of a *nonlinear* entity [3]. It indicates that man and his means form the functional system together and cannot be defined without each other. I have to remark, that it is the observer who, for better understanding, constructs the structure of a phenomenon. For a natural entity itself structure has no meaning, since, being divided into pieces or disconnected, the entity loses its very nature of wholeness and totality.

The basic feature of all complex functional systems, next to the manifestation of evolution, is their *self-preservation*, *homeostasis*, maintenance of the undisturbed algorithmic operation of its vital processes. The track guidance of *inner algorithms* (memory, motivation, sensory control, etc.) and *outer influences* (impacts of hierarchy, form, and mechanisms; interactions between systems; etc.) help the system to keep the trails of the past and allow to replicate itself periodically *without turning into chaotic oscillations*. This mode of quasi-stable, self-preserving operation called *homeostasis* or *homeokinesis* does not exclude qualitative changes but keeps them within limits, preferring self-preservation of function on the expense of slight changes of structure.

Even the simple reiteration of proficient performance leads to a step by step, but steady, increase of competence, experience and efficiency. But the real qualitative change can be reached by implementing modulation correction. The functional system is capable not only of reiteration of its activity. A drastic change may also be accomplished to modify its own function seriously. Therefore, in case of socio-technical functional systems two, extreme but complementary of each other, patterns of change can be distinguished:

- a *paradigm of stability* or routine replication, and
- a paradigm of modulation correction or *creative self-modification.*

Routine is an indisputable dominant paradigm in society, since the slow rate of adaptation makes its control safe and possible. The paradigm of creation is, on the other hand, in sharp irreconcilable controversy to the routine, since it jeopardizes the stability and structural integrity of the whole hierarchy for the sake of qualitative modification of its performance. It is creation that changes the dominant routine and gives to it the context of activity. The being and becoming, the existence (as adaptation by moderate modification) and change (as qualitative self-modification), the routine and creation are the patterns or phases of functioning a system distinguishes.

The functional system periodically re-estates itself in a sequence of qualitatively different phases, and may therefore well be called a moving or *evolving system.* The evolutionary movement of the system is characterized by meta- or dynamic stability, stable if in movement. Evolution means that an evolving functional system can therefore either be in the process of its progression (sequential phases tending toward *qualitatively higher evolutionary levels*) or in retrogression (movement toward *degeneration*). From the system's point of view, progression is a steadily growing *self-developing* ability, the way of self-perfection. Retrogression, on the other hand, is a steady decline of its performance, a determined reduction of the problem-setting and problem-solving ability, which inevitably leads to gradual narrowing of the system's scale of activity, the *specialization*, sacrificing its own parts, and finally to its disintegration.

The mechanism, as an exertion of outer impact, guides the evolutionary movement of a socio-technical system as a stabilizing control block. Therefore a '*system-mechanism*' coupling determines as a constraint the evolution of the otherwise undetermined individual functional system. Anything that exerts a kind of external influence on the functional system serves to it as a mechanism. The firm, the institutions, the symbols of communication, instruments and machinery can be mentioned as typical examples of social mechanisms. In this respect, for socio-technical systems the whole economy appears as a set of different mechanisms.

Two specific basic functional couplings of interacting socio-technical systems can be distinguished: cooperation and competition. *Cooperation* means a *positive amplifying feedback*, systems attending each other, one's result serving as the other's tool. *Competition* means a *negative regulating feedback*, a challenge and searching for new solutions. The external mechanism connects these interacting systems to each other like a gearless transmission into a specific steady '*cooperation-competition*' formation, when system A cooperates with B, B with C, but C competes with A. These couplings are the 'long lasting transient' systemic formations of the human society and can be distinguished as functional macrons.

A socio-technical system usually does not interact with nature as such but rather with other socio-technical systems. Those systems, their out-of-action components, mechanisms and stand-by instrumentality form a kind of specific functional environment around each socio-technical system. The interaction of the [*Human Being and his Means*] system with its functional environment principally differs from the interaction with nature, and can acquire either cooperative or competitive character. Cooperative feedback from functional environment can multiply the system's ability and its performance. Competitive feedback forces the system to adapt itself to the environmental functional changes and to find new solutions to its problems. To a socio-technical functional system nature means just a conglomerate of self-preserving homeostatic systems, albeit the basis and condition of its very existence. But since the [*Human Being and his Means*] system and its functional environment may produce a multiplying interaction, an exciting impetus of modulation correction, this can principally lead to *uncontrollable oscillations* and—in an extreme case—to a catastrophe, breaking down the always limited natural self-restoration ability.

The higher the number of out-of-action components and stand-by means from which the socio-technical functional system is mobilizing itself from, the higher the degree of its success and more steady its evolution. The scale of plasticity of the functional environment, the number of the degrees of freedom of would-be components, and the number of alternatives the system is able to anticipate and maintain, and therefore its *freedom and control*, are in strong correlation with the rate and level of evolution. The *law of redundancy* means the necessity of growth of the functional environment as a precondition of socio-technical functioning on the *ever higher level of evolution.*

The levels or *stages of cognitive evolution*, the dynamic structure of its process can be described by models of *Piaget* and *Ponomariov* on the basis of their experimental investigation of the development of the individual's mental capacity. The stages of development of the intellect constitute active structural levels of the integrative organization of problem-solving ability from intuitive to logical [13]. In the case of society we take the individual's cognitive change stochastically, therefore these stages or levels may be conceived as structure models of evolution of human intelligence.

CONTRADICTIONS AND RISK-FACTORS

The list of risk-factors we are aware of (courtesy of the late Dr. József Horváth, University of Lóránd Eötvös, Budapest), the worry we feel about the corporate and global progress of our days is an expression of our apprehensiveness on a higher level of development:

Natural risk-factors:

1. Cosmic catastrophes
2. Geological catastrophes
3. Biological catastrophes

Social risk-factors:

4. Over-population and inadequacy of resources
5. Destruction of living-space, environmental pollution
6. Wars
7. Crime and violence
8. Aberration of life functions (drugs, etc.)
9. Global economic differences
10. Unrestrained development of science and technology
11. Crisis of restructuring in industrially developed countries
12. Conflicts among civilizations, ethical dissimilarities

Mental risk-factors:

13. Nationalism, fundamentalism
14. Moral degeneration
15. Emotional and mental aberration

The tendency one may notice in this list of risk-factors points to the distress and suspicion most of us feel in case of any change. The *change*, and therefore development, is always accompanied by contradictions and a certain amount of *risk*. Whatever the human being has created, has always been meant to solve the contradictions created by himself. Therefore, most of the risk-factors mentioned above are of this conventional character or its modern manifestation and belong to one of following categories:

- *Functional disturbances* due to, for example, the principle of resource and time allocation;
- *Contradiction of 'Part' and 'Whole'*;
- *Contradiction between homeostasis and modulation correction.*

If the system, or the hierarchy it belongs to, cannot 'digest' the major disturbing influence and the contemporary socio-economic mechanism is never good enough to bridge it, the contradiction may be called antagonistic. However, disfunction and disharmony will occur, if the contradiction cannot be removed as a disturbing factor. In this case the system may disintegrate, but its disintegration will also painfully influence the connected functional systems. Via the functional system above it affects the whole *hierarchy* as well. Disfunction of the system would at the same time also destroy, or condemn to disintegration, the subsystems at a lower level in the hierarchy. The prolonged, 'irreconcilable' local *contradiction* thus may become a *general crisis*, a factor impeding development of the whole hierarchy. Antagonistic contradictions might lead to *catastrophe* if their scale exceeds certain limits and the catastrophe already cannot be prevented, the contradiction cannot be solved yet. The antagonistic contradiction cannot be eliminated as such, since it is a part of the process, but a creative intellect may overcome the obstacle, providing a solution, a witty socio-economic mechanism to bridge the contradiction.

The mental mode of life, the cognitive evolution brought up on the higher level, didn't put an end to the functioning on the lower levels of evolution. On the contrary, mankind has had to become acquainted with intellectual forms of underdevelopment, like mental crime or highly intellectual incapability. Risk-factors 12-15 indicate that this intellectual or moral conflict may manifest itself on national level or world-scale as well.

Restructuring

The *world-wide computerization* brings about the imperative of restructuring in industrially developed countries. The walls of organizations became transparent as far as the flow of information is concerned. The flood of symbol-capital around the globe nowadays exceeds the volume of the trade turnover almost 60 times. The *multinational economy* gradually places itself above the state or national control, and that aggravates global economic differences, jeopardizes the well-fair societies (investment into the future of the society; equalization of social differences). This widely brings about not only a feeling of unrestrained development of science and technology, but foreshadows a *political crisis* of unknown magnitude.

Limits to Resources

Limited resources may become a barrier if it happens unexpectedly and its effect manifests itself sooner than the identification and solution of the problem. But for cognitive functioning the lack of resources or the contradiction between homeostasis and modulation correction just formulates the problem of *substitution* or any other way of self- and mechanism-adaptation to changing conditions. The trend of energy consumption, for example, represents the rate of socio-technical development of the human race: till 1850 humanity consumed approximately 6-9 Q of energy resources (where $Q=10^{18}$ BTU and BTU—British Thermal Unit—is equivalent of $26 \cdot 10^{12}$ tons of crude oil), and 5 Q from 1850 till 1960, when the world population doubled from 2.4 to 4.8 billions. For the 21st century the necessity of 100 Q of energy is forecasted. For us it means that new energy resources ought to be found, otherwise the rate of socio-technical evolution will slow down. It is not yet a barrier which brings the process of evolution to a halt but a serious warning.

Boundaries of Natural Self-Preservation Ability

On the other hand, cognitive evolution brings its own rules into this process of self-adjustment and adaptation and shakes its balance. The law of redundancy here means that the expansion of socio-technical functional environment should overwhelm the rate of functional change of systems to counterbalance the inevitably generated 'noise' (useless but to-be-processed information) and to provide a greater variety of possible means and solutions to make the rather uncertain cognitive and socio-technical positive change more probable. During this process the natural environment around us—a conglomerate of self-preserving nonlinear homeostatic systems—is being rapidly substituted by an artificial, linear, anthropomorphic environment, which endangers the very self-preserving ability of natural systems. The socio-technical form of cognitive evolution principally means for us permanent and determined substitution of natural environment by modified and linearized one. It is like the process of *cancer development* in an organism, replacement of original tissue by degenerated one. If uncontrolled, the continuous positive feedback from the artificial environment to the functioning system may cause unexpected oscillations and unknown catastrophe. The first sign of injury of nature, we may notice, will be most probably the unexpected disorder of our own self-restoration ability, the disfunctioning of our own immune system. This built-in antagonistic controversy means a serious barrier on the road of cognitive evolution. The problem is not, for example, the quantity of energy we need in the next century, but the predictable break down of self-preserving capability of the Earth, when byproducts of the consumption of 100 Q of energy are discharged in its atmosphere. What we know for sure is that self-preservation of any homeostatic system has its own limits.

Weapons of Mass Destruction

Another example, where a functional contradiction becomes a barrier, is when the conflict occurs between two competing functional systems, which differ from each other, or—having common components—partly overlap. If the two systems are members of different hierarchies, and the hierarchies do not submit themselves to the jurisdiction of the common center above them, the contradiction may become antagonistic. Think, for example, about a war between two countries or a threat on behalf of a self-mesmerized organization with weapons of mass destruction in its hands. The contradiction can, however, be bridged if it assumes the form of the conflict of 'Whole-Part'.

Evolutionary Gap

The gap between the highest evolutionary level achieved and the lowest one in the hierarchy is inevitably enlarging in the course of cognitive evolution, even if the lower levels are modified, significantly lifted up by the impact of higher levels. It is not simply the tension or the resistance and aggressive conservatism of the majority, as a common controversy. In the long run it means an irreconcilable and antagonistic contradiction between *rapid cognitive evolution and hardly changing mental routine*, which can easily turn into contradiction between the progress and retrogression. If the crisis of retrogression takes place, *the bulk of society* simply descends to a lower level of evolution, less significant and more backward than it was formerly. This is the crisis of cognitive development or permanent intellectual creation, the breakdown of the 'Idea → Implementation of knowledge → Education → Thought' functional macron of society. The first sign of such evolutionary retrogression is *injury of ethical values* that can be followed by degradation of education.

The critical point of mental retrogression is the intellect's inability to recognize the process of retrogression and to mobilize all the social will-power with the super-ordained aim of self-confidence or the principle of highest moral responsibility. Mental retrogression accompanied by political disequilibrium, aggravating economic weakness, backwardness in education and culture, deteriorated by centralistic hierarchy and the spasm of *ideological fundamentalism and intolerance* is bound to bring the society to the lowest level of mere physical survival, the nightmare of self-strangling.

It is first of all education that may encounter its natural physiological boundaries. For most of us, the mental way of life already means a practically unceasing, life-time learning. The *educational bottle-neck* of human cognitive evolution, the methods we use in education hardly satisfies even today's requirements and, therefore, cannot meet the higher necessities of tomorrow.

The Trap of Logic

Thanks to our ability to analyze a given natural phenomenon and synthesize knowledge from crumbs of our fragmented observations, mankind was able to rise to the heights of its socio-technical evolution. Thanks to well defined rules of logic, we are able to forecast predominant, albeit mostly linear trends, to adjust our plans of activity to each other, to arrange our knowledge into a linear but hierarchical and therefore mechanically processable arrangement, and to transfer it together with algorithms of logical argumentation to the next generation. The *art of logic*, the perception of the world through postulating its *linear causality*, is our strength. And logic is our weakness, since *most of natural functional phenomena are of nonlinear character.*

We do not normally master the totality as such. Only some of us are able to grasp wholeness *by intuition.* Alas, training of human intellect by logic, so needed for our development, makes even those few mostly 'totality-blind'.

Trying to grasp the truth of nonlinear reality, but developing our intellects by means and rules of logic, we condemn ourselves to the asymptotic approach to reality without the hope of ever reaching the truth or 'absolute' knowledge, the ability of clear vision of natural phenomena. Cognitive development by means of logic theoretically may create for us antagonistic contradictions on high levels of evolution, since approaching asymptotically the reality or the axle of knowledge, our efforts will require exponentially growing efforts and energy.

The '*totality blindness*' is already in front of us as a barrier of scientific progression on the route of cognitive evolution. In the long run it is our consciousness, rather than intellect alone, that we shall have to trust and rely on. Human consciousness with its intuitive abilities, supported by a trained intellect, may well become the subject matter of human evolution directed toward the ideal of self-development. Let's say, at that stage of human development the ideal of cognitive evolution will most probably be replaced by the *ideal of evolution of Gnostic consciousness.*

PRINCIPLES OF POLITICS

The more we know about our evolution the higher is our responsibility. The responsibility we bear means new principles of ethics and philosophy, as well as institutionalized background support for their implementation.

It is already clear that such principles as 'protection of environment', 'restricted or optimum intervention', 'preservation of the level achieved', 'minimal caused damage', 'restoration of damage', are not sufficient to satisfy our responsibility for the rate and level of cognitive evolution we have managed to realize. Active policy requires *creation of favorable conditions for individual self-development as well as for community progress.* It appeals for active cultivation of a plentiful and multi-colored functional as well as natural quality environment, and not just its passive protection. The principle of civil *rights* for cognitive evolution should say that *everybody is entitled to the qualitative natural environment*, as well as to the qualitatively *appropriate socio-technical one*, which optimally protects, serves and stimulates the self-development of the individual.

The principle of community *rights* for cognitive evolution should say that the community that feels *responsibility for its mental way of life* and cognitive evolution is entitled to *social support and assistance* unless its activity creates artificial limits for individuals and forces them to restrict their self-development.

The concept of *crime against human evolution* should lead to the principle of individual responsibility for social cognitive evolution, and, vice versa, social responsibility for individual evolution, that should emphasize the need of active encouragement and support on individual, community, as well as social level.

The principle of functional environmental redundancy or abundance of alternatives is derived from the fact that the greater the variety of alternatives a functional system may select from, the more smooth and far reaching is its self-development. The structural conservatism, which manifests itself inevitably in any hierarchy, could and should be weakened by special socio-economic mechanisms and germs of its negation.

The most conservative part of any social institutionalized structure is its control function, which has to perceive and signal any deviation from the ever fundamental unquestionable pattern, the very routine cognitive evolutionary movement is bound to change. The application of the principle of functional environmental redundancy or abundance of alternatives should mean in this respect a greater tolerance on behalf of control institutions. The conflict between the monopolistic hierarchical control and the avant-garde of socio-technical evolution seems to be of permanent character and should be counterbalanced by means and concepts of policy. The same stands for minimizing and sharing of the risk, any change is accompanied by.

Technology policy is the policy of promotion of self-development and offensive strategic creation of technological advantage economically utilized on the *world market*. It is world, national or regional technology policy that may help to keep and stabilize the pace and direction of socio-technical progress of the region, paving the road of cooperation between its socio-technical actors, aiming the sensors of institutionalized control to their *innovation performance*. It is technology policy that should govern the assessment of new technologies, monitoring their social effect, controlling and handling their critical and destabilizing impact on the mode of life society is accustomed to. It is technology policy that is responsible for coordinating international cooperation, preventing the country's marginalization on the world market.

The principle of responsibility for human cognitive evolution is even more important and serious since mankind is facing the drastic qualitative change of the very pattern of its development. We are not aware of specific characteristics and accordingly the pitfalls of the extreme change we are moving toward, nor its speed and shape. To turn 'en masse' to cognitive evolution took mankind 400 year, from Gutenberg to mass education at the second part of the last century, and 100 years, from almost complete elimination of illiteracy to the information world-society with its mental mode of life. At its last stage cognitive evolution was marked by rapid development of technology, economic growth and crises of unknown magnitude. Were we capable to recognize in time the essence of evolutionary changes, we could have been able to find proper socio-economic mechanisms to avoid and prevent all the 'unavoidable' contradictions, unexpected oscillations and catastrophes we have had to face. The price we paid for the cognitive pattern and higher level of our evolution was very high. I believe, humanity has now sufficiently grown up to take all the necessary precautions. At the very gate of alteration from the pattern of cognitive evolution to the pattern of evolution of Gnostic consciousness this is the task humankind has to formulate for itself in order to find solutions and proper socio-economic mechanisms for this arduous phase-transition.

Borisz Szántó
Academic Research Unit
Technical University
Budapest, Hungary

REFERENCES AND LITERATURE

[1] **Anokhin, P.K.**, *Izbrannye trudy. Sistemnye mehanizmy vysshej nervnoj deiatelnosti (Selected Works. System Mechanisms of the Highest Nervous Activity)*, In Russian, Nauka, Moscow, 1979.
[2] **Anokhin, P.K.**, *Uzlovye voprosy teorii funkcionalnoj sistemy (The Key Questions of the Theory of Functional System)*, In Russian, Nauka, Moscow, 1980.
[3] **Biro, V.**, *Nonlinear Oscillations in Feedback Systems*, Akademiai kiado, Budapest, 1985, p. 7.
[4] **Crook, John H.**, *The Evolution of Human Consciousness*, Clarendon Press, Oxford, 1980, p. 36.
[5] **Gardner, Howard**, *The Quest for Mind*, Alfred A. Knopf, New York, 1973, p. 236.
[6] **Haken, H. & Wagner, M.** (eds.), *Cooperative Phenomena*, Springer Verlag, Berlin, 1973.
[7] **Haken, H.** (ed.), *Evolution of Order and Chaos in Physics, Chemistry, and Biology*, Springer-Verlag, Berlin, 1982.
[8] **Jantsch, Eric**, *For a Science of Men*, in: **Urban, G.R.** (ed.), *Can We Survive Our Future?*, The Bodley Head, Ltd., London, 1972.
[9] **Jantsch, Eric**, *Evolution: Self-Realization through Self-Transcendence*, in: **Jantsch, E. & Waddington, C.H.** (eds.), *Evolution and Consciousness. Human Systems in Transition*, Addison-Wesley Publishing Co., Reading, Massachusetts, 1976.
[10] **Piaget, Jean**, *La psychologie de l'intelligence*, Armand Collin, Paris, 1967.
[11] **Piaget, Jean**, *Psychology and Epistemology. Toward a Theory of Knowledge*, Harmond Worth, Penguin Univ. Books, London, 1972.
[12] **Piaget, Jean**, *Afterthought*, in: **Koestler, Arthur & Smythies, J.R.** (eds.), *The Alpbach Symposium 1968. Beyond Reductionism. New Perspectives in the Life Sciences*, Hutchinson, London, 1969, p. 158.
[13] **Ponomariov, G.A.**, *Psichologia tvorchestva (Psychology of Creativity)*, in Russian, Nauka, Moscow, 1976, p. 32.

[14] **Prigogine, I., Stengers, I.**, *Order out of Chaos. Man's New Dialogue with Nature*, Bantam Books, New York, 1984.
[15] **Szántó, Borisz**, *A teremto technologia. A tarsadalmi-technikai evolucio elmelete (Creative Technology. The Theory of Sociotechnical Evolution)*, in Hungarian, Kozgazdasagi es jogi konyvkiado, Budapest, 1990.
[16] **Vernadski, V.I.**, *Razmyshleniia naturalista. Nauchnaia mysl kak planetnoe iavlenie (Meditations of a Naturalist. The Scientific Thought as Global Phenomenon)*, in Russian, Nauka, Moscow, 1977.
[17] **Whitehead, A.N.**, *Adventures of Ideas*, MacMillan Publishing Co., New York, 1967.

DMITRI SPIVAK

ALTERED STATES OF SOCIETY: A TENTATIVE APPROACH

INTRODUCTION

The plausibility of transferring the notion of 'altered states of consciousness' to social sciences is discussed in the framework of problems and new trends of the post-communist society in the East, compared to the post-industrial one in the West. Transferring is regarded as possible to the extent to which sociocultural groups coincide with subcultures of altered states of consciousness in the given society. The article at hand is based on a first-hand account of current discussions in Russian academic circles.

Ars longa - vita brevis - occasio praeceps - experientia fallax - judicium difficile, Hippocrates says, "science is long, life is short, opportunity is transient, experience is delusive, and judgement is difficult". The founder of medicine primarily aimed at impediments on the way of professional treating of the human organism. However, these impediments could be attributed to the progress of those having to do with the flaws of the social organism as well.

The objective of the present paper consists in discussing the plausibility of transferring to social sciences some essential notions and description procedures of the theory of altered states of consciousness, elaborated currently in the domain of psychology.

The general framework of this theoretical move is formed by rapidly evolving features of the New World Disorder, to use a felicitous expression coined in the recent book by Z. Brzezinski (and several other authors, [5]). We would begin with three sets of problems presenting briefly some symptoms of landslide changes occurring currently in the collective psyche.

Fear

Fits of unmotivated fear in everyday life are characteristic for about 9 Russians out of 10, according to a recent study done by the Center for Social Prognosis and Marketing. There is some dynamics about this problem, which does not seem quite essential: fits of the feeling of anxi-

© 1999 *Kluwer Academic Publishers. Printed in The Netherlands.*

ety, crisis or catastrophe were admitted by 94-95-93-91-89% of Russians studied by the Center in the years 1990 through 1994 [17a].

Trying to rationalize fears is proper for both the population and the scientific community. Consequently the former rush one time to buy goods which could go up in price due to rapid inflation, another time to join courses of English (a popular Russian parable says that it is optimists who study English, while pessimists study Chinese, and realists study the Kalashnikov sub-machine gun). As to the scientists, models of factors conditioning mass fear and frustration in general are being constructed.

The set of factors seems to be quite stable, while the ratings can differ. To cite a serious mass study, people are currently afraid of 'criminality-rise in prices-disintegration of the country'; according to another one, the objects of fear are civil war-criminality-poverty-unemployment (the sequence of factors is cited in both cases in the descending order, cf. *Phenomenon of elections in Russia* [2].

The structure of such lists seems to be conditioned by both rational and irrational factors [24]. This could be demonstrated on the example of the last sequence cited. The fear of poverty rose from 28 in 1992 to 42% in 1994; the fear of civil war went down from 62 to 45%, at the same period of time. The former process was quite natural, reflecting the failure of the market reforms; the latter seems quite irrational, taking into account skirmishes in Moscow (1993), and the impeding Chechen war (1994). It is the combination of such factors that tends to make interpretation and prognosis fuzzy.

Apathy

Unmotivated apathy as the constant or recurring background of one's everyday professional and personal life is characteristic for about 4 Russians out of 10 [22]. Researchers having to do with this and related phenomena, tend to emphasize its paradoxical character. For a substantial example, data of a mass study of qualified Russian workers, obtained by the National Center for Study of Public Opinion, are cited below [11].

The majority of the workers neither obtain sufficient salary, nor have other sources of income (66%). They regard themselves as able to cope with a much more intense work (66%), and do not in principle reject free market economy (60%). Thus there seems to exist no major inner hindrance to their productive labor.

This image is contradicted by the fact that about half of the workers are ready to renounce high but relatively insecure salary in favor of a low but stable one; and to leave overtime labor in favor of rhythmic and slow one (the exact level is 48%, rising to 64% for unskilled workers).

There is quite a lot of reliable data of this kind. They tend to be interpreted usually as revealing the inclination of workers to some kind of paternalist model, in the framework of the free market system. This thesis is corroborated by independent studies of the political orientation of both workers and managers. When asked the corresponding question directly, both groups definitely preferred the paternalist (and/or social democratic) model to the liberal one: in fact, the ratio was 2:1 [3].

At the same time, the Russian government keeps firmly professing liberal ideals, and implementing them in its practical activity; hence the imminent alienation. The general population seems to be in this respect on the side of the workers: about 80% have acknowledged feeling distrust in the government; not less than 60% would definitely prefer to have a very strong president. It has been supposed that this conflict of socio-economic values could result in rapid re-orientation of Russia from the American model to the Japanese one, with obvious geopolitical consequences [18].

The basis of social alienation could thus seem to be quite rational. Consequently, it would have been natural for the workers, feeling an inclination to paternalism, to support institutions that would be able to introduce it. However, about 87% of workers have acknowledged mistrust in trade unions [11].

This is quite strange: trade unions are firmly associated in Russia with the old Communist paternalism, which retains its attractiveness for at least half of the workers. The situation is even worse for the political parties: they are mistrusted by more than 90% of the population, although a number of them have included paternalist issues into their programs. To complete the picture, the only object of trust of a considerable part (40%) of the population seems to be the army, although it is definitely unable to introduce any socio-economic model whatsoever [17b]. It is in bringing together such contradicting values that unmotivated apathy would be the most plausible outcome to arise.

Mysticism

Non-traditional, primarily mystical spiritual orientation is currently characteristic for not less than 1 to 4 Russians out of 10, depending on social stratum and other secondary factors (data of a recent survey by the Analytical Center of the Russian Academy of Sciences [7].

The outline of the group seems quite fuzzy, ranging from persons who either believe in super-human powers, or like to read mystical literature, and all the way to those regularly taking part in occult ceremonies.

The group has only recently been shown by a separate entry in Russian statistical surveys, and quite in time: there is enough evidence to suppose that it has played a key role in bringing democracy to Russia. We would discuss in some detail three of its aspects, forming a necessary background for our topic.

First, the 'mystical' group seems to be superimposed upon the traditional division into believers, atheists, and those hesitating or largely indifferent to religion. One could regard it as belonging to a separate dimension, if it would not now and then compete with religion.

For instance, the share of Orthodox Christians in the population of Russia went down from 40 to 19% in the years 1990-1991. This process has reflected landslide changes in the religious life of the country, since the share of Orthodox Christians in the bigger group of Russian citizens belonging to some religion was at the level of about 90%. Where did these dissidents take way? Simultaneously, the share of 'Christians in general', belonging to no definite Christian denomination at all, has risen from 22% to 47%. It would be quite possible that a large part of the dissidents came right here.

Second, the 'mystical' group tends to consolidate and reveal much more activity than usual on turning-points of history. The activity is enhanced by its young age: approximately 35% of Russian teenagers (aged 16-17) are mystically oriented (in people aged 60 and older, the corresponding figure is about 9%). As to consolidation, it is enhanced by the process of modernization in adjacent realms: e.g., the major statistical surveys have demonstrated Orthodox creed to have recently become a religion of young educated urban dwellers (in contrast to the previous generation, which had tended to be aged, less educated peasants by birth).

Third, the 'mystical' group has revealed a tendency to advocate not religious, but political slogans, or rather to join them. For instance, it was about 62% of mystically oriented citizens who advocated the values of perestroika in 1991, while 26% belonged to the opposite side. About 40% of persons believing in astrology, 44% of adherents of Oriental wisdom, and 44% believers in UFO, acknowledged themselves as active supporters of the Western values. There are good reasons to suppose that this proportion is fairly representative for the mass movement that brought Russian reformers to power in 1991.

Based on observations of this kind, there were columnists in 1991 who tentatively depicted the mental landscape of a Russian democrat as a place where self-styled yogis nurtured their Kundalinis, Don Juan instructed Carlos Castaneda in the mysteries of nagwal, and UFOs glided around under a big rainbow.

However it is this type of factors that shapes the political life of Russia nowadays. One of its legs is formed by fundamentalists of all possible creeds, from the Orthodox to the Communist ones, while on the other one modernists and post-modernists, coming from different sides, are concentrated.

It is in such matters that some obscure steps of President Yeltzin find their explanation. One could cite here his personal admittance of the order of the ancient Assyrian goddess Baw, rewarded by a leading spiritual healer, Juna Davitashvili. Journalists present at the ceremony felt somewhat puzzled; however, serious analysts acknowledged the President to feel the attitudes of his electorate.

One of the specific features of a considerable part of it is predilection for weak political power. Any attempt of the President to introduce elements of a strong or authoritarian rule would result in the transition of this group to another destabilizing leader. This could have contributed to the result of the notorious elections of 1993 when a considerable part of the Russian democratic electorate changed their minds, quite unexpectedly for the leading political analysts. However, this is a topic for a special work; as to the present one, suffice it to state that an irrational factor has penetrated into the very core of present-day Russian politics.

EAST AND WEST

The list of factors is by no means complete; whether it is finite remains unknown. The factors cited look rather like secondary symptoms, and are most probably heterogeneous. Thus a specialist in religious studies would duly remark that the factor 'Mysticism' should be regarded as to a large extent conditioned by factors 'Fear' and 'Apathy'.

However, the general diagnosis is definite: it consists in a sharp crisis of the society, affecting its very foundations. P. Bourdieu would speak about a flaw of the entire realm of social topology, including the symbolic sphere; R. May would state the onset of mythoclasm, i.e., the wreck of basic social myths; K. Popper would suppose a massive crisis of social institutions.

The post-Soviet crisis being marked by peculiar traits, similar processes may be discerned in the West as well. Thus American workers may appear dynamic compared to their Russian counterparts; however they would appear quite passive compared to Japanese or Korean workers. As to immediate causes, a wide range could be cited in this respect, from the Western system of social benefits—to the national stereotypes of labor discipline (factor 'Fear').

Waves of irrational fear have also been registered in the Western society. They affect periodically not less than 1 German out of 10, which numbers about 8-9 million persons. Approximately 1 million out of this number experience 'Fear' as the constant background of their lives; about 6% of the population of Germany have at least once consulted their physicians in connection with this state [4]. The range of possible causes cited in this respect in scientific literature tends to be rather broad, including both social and biological factors, which presupposes the necessity of an interdisciplinary approach, cf. Interdisciplinary fear [1] (factor 'Apathy').

The onset of factor 'Mysticism' could most plausibly be connected to the growth of New Age spirituality, or new religious consciousness in general. The Western world reveals a distinct process of this kind. About 2 out of 3 citizens of the EC countries regard themselves as religious, but not belonging to any specific denomination (although they may acknowledge adherence to the Christian heritage in general [8]).

However, special studies tend to state that the roots of the new type of spirituality in Russia do not fully coincide with those in the West. To cite only one study, the 'old' traditional mysticism has tended to play a dominating role in the former case, even compared to the side effects of modernization; in the West, the proportion may be regarded as reverse [16].

To sum up,the factors discussed above seem to be to a certain extent parallel and sometimes similar in the post-industrial West and the post-communist East, but definitely not identical. However, both facets seem necessary for a consistent description and prognosis. To do this, one would have to supplement the traditional apparatus of description of crises and modernization processes, well-established in social sciences, by notions oriented specifically to apparently non-motivated, fuzzy and irrational factors.

ALTERED STATES OF CONSCIOUSNESS

The concept of altered states of consciousness, elaborated in contemporary psychology, seems to provide useful insights into our topic. A well-known anthology edited by Ch.T. Tart [20] (first edition 1969) remains the standard reference work.

Development prior to this publication was mostly latent; specialists in the history of psychology have traced it back as far as to W. James and W. Wundt. Development after this date was too impetuous and multifaceted to be generalized without a special study, which has not yet fully been done [12]. For the sake of brevity, we would attempt to define the concept of altered states of consciousness in a few tentative points:

- *intuition of clusters*: consciousness is structured not as a homogeneous phenomenon or process, but rather as a cluster of states, each one of which is governed by qualitatively different regularities;
- *intuition of focal points*: consciousness functions as a ray, focusing on a chosen state, or a finite number of them. A sequence of such foci is fixed for a given culture, forming the basis of habitual everyday consciousness, regarded as non-altered;
- *intuition of shifts*: the sequence of focal points can shift in the space of consciousness, primarily for adaptive purposes. In this way conscious activity under considerable strain, caused by both outer and inner factors, is provided.
- *intuition of transcendence*: cases of involuntary "sticking" in a state of consciousness, or between such states, as well as trips to remote parts of the field of consciousness, tend to produce a wide range of subjective experiences, often including such ones as fear, apathy and mystical feelings [13]. Similar experiences may be voluntarily induçed for personal growth or creative purposes, in the framework of a psychotherapeutic process. This is why labeling them as positive or negative depends rather upon outer values and expectations. Association to factors 'Fear-Mysticism' cited above, would in principle be justified.

SUBCULTURES OF ALTERED CONSCIOUSNESS

Direction of further inquiry seems evident. One has to transfer the notion of altered states of consciousness upon the level of society, to transcript its intuitions into sets of categories and methodologies, and to obtain finally a novel approach to mass consciousness, and social politics. In doing so, one would follow a general contemporary tendency towards psychologization of social sciences.

In general, the plausibility of such a move (transferability, to use the term of I. Stengers) has to be demonstrated in each case. In our case, it is to be approached basing on the results of contemporary ethnology. Having studied a representative sample of traditional societies, E. Bourguignon stated that altered states of consciousness were acknowledged and institutionalized in about 90% of them [23].

The types of institutions tended to differ greatly. However, their main function remained handling the flow of experiences generated by trips into altered states of consciousness, in order to reduce their destructive component, and to reshape the rest for the benefit of the society [9] [10]. The institutions implementing this function may be defined as subcultures of altered states of consciousness. At this point, the level of individual psychology is to be supplemented by that of the interpersonal relations, and to a certain extent of societal bonds.

To delineate their combination, one would have to trace back the long history of such subcultures, providing first a sacred dimension in the traditional cultures, later a background for passage from the community to the society (*Gemeinschaft* vs. *Gesellschaft*, to speak in terms of F. Tonnies), and finally to be split and divided by a single process, presented in contemporary theology by an impressing sequence: 'industrialization, urbanization, technologization, bureaucratization, scientism, instrumental rationalization, secularization, egalitarianism, and materialism' [6a] [6b].

SOCIOCULTURAL GROUPS

Sociologists have themselves been interested in the behavior of Man and Society in the Age of Crisis. We use capital letters to mark the title of a well-known book by the founder of sociodynamics P. Sorokin, published in 1941. Sorokin's conclusions are representative for modern sociological thought, and reveal quite definite affinity to the transpersonal paradigm. We would reformulate them in terms of intuitions, referring to corresponding entries introduced above:

- *intuition of clusters*: society comprises a loosely structured multitude of sociocultural groups, embodying its variety and complexity;
- *intuition of focal points*: an individual potentially belongs to a set of such groups, and actually passes them in the process of life and purposeful activity, following a route acknowledged or tolerated by the society;
- *intuition of shifts*: passages from one sociocultural group to another are inevitable, and comprise the background of one's life;
- *intuition of transcendence*: to manage the crisis, one has to carry out a restructuring of personal value systems, ego sets, and patterns of group connections. One of the ways to achieve it is their deep disintegration, followed by reintegration along a new pattern, which may be regarded as altered [19].

The level of sociocultural groups may be regarded as the lowest one fully conforming to the laws of society in general, decisive for the processes of its crisis and reintegration, and at the same time implying to a considerable extent the notions belonging to levels of interpersonal relations and individual psychology. Subsequent innovations have introduced numerous amendments into the understanding of sociocultural groups, without erasing this important trait [14a] [14b] [15].

CONCLUSION

Transfering the notion of altered states of consciousness upon the level of society is methodologically plausible to the extent, to which subcultures of altered states of consciousness coincide with sociocultural groups in the given society. This conclusion, theoretically introducing the notion of altered states of society, forms the main theoretical result of the present paper; its political (and epistemological) consequences are to be elaborated in a special work.

ACKNOWLEDGEMENTS

The author is grateful to Ms. K. Wistrand for her constant and generous support, and to Dr. A. Kolman and Dr. F. Janouch for their help in producing the final version of the manuscript.
This work was supported by grant # 98-06-08004 from the Russian Humanitarian Scientific Foundation.

Dr. Dmitri Spivak
Human Brain Institute
Russian Academy of Sciences
St. Petersburg, Russia

REFERENCES

[1] **Anonymous**, "Interdisciplinary fear", *Chelovek*, **2**, 1994, pp. 145-146 and 152-153. (*)
[2] **Anonymous**, "Phenomenon of elections in Russia: socio-psychological analysis (round table discussion)", *Psikhologicheskij Zhurnal*, **4**, 1994, p. 97. (*)
Problems of Russian geopolitics, Vestnik MGU / Vyp.12, **6**, 1994, pp. 3-22. (*)
[3] **Avraamova, E. & I. Diskin**, "Social transformations and elites", *Obshestvenny je nauki i sovremennost*, **3**, 1994, pp. 14-26. (*)
[4] **Bovkun, E.**, "Fear: new epidemic of our times", *Izvestia-Digest*, **7**, 29.09.1994. (*)
[5] **Drucker, P.**, *Post-capitalist society*, NY, Harper Business, 1993, p. 113.
[6a] **Griffin, D.R.** (ed.), *Spirituality and society: postmodern visions*, State University of NY press, Albany, 1988, p. 9.
[6b] *ibid.*, p. 12.
[7] **Furman, D.**, "Religion and politics in post-communist Russia (round table discussion)", *Voprosy Filosofii*, **7**, 1992, pp. 13-15. (*)

[8] **Hjorth, C.**, "Pa jakt efter den europeiska sjalen", *Dagens Nyheter*, 26.03.1995, A8.
[9] **Katz, S.**, *Mysticism and philosophical analysis*, Oxford University Press, Oxford, 1978.
[10] **Kimelev, J.**, "Phenomenon of mysticism and contemporary bourgeois philosophy of religion", *Voprosy filosofii*, **8**, 1985, pp. 140-148. (*)
[11] **Komarovskij, V.**, "Transitional consciousness of a transitional period", *Obshestvennyje nauki i sovremennost*, **1**, 1994, p. 41. (*)
[12] **Lajoie, D. & Shapiro, S.**, "Definitions of transpersonal psychology: the first twenty-three years", *The Journal of transpersonal psychology*, **1**, 1992, pp. 78-98.
[13] **Ludwig, A.**, *Altered states of consciousness.* In: **Tart, Ch.T.** (ed.), *Altered states of consciousness*, Anchor Books, Garden City, 1972, p.16.
[14a] **Matveeva, S.** (ed.), *Modernization in Russia and conflict of values*, Instit ut filosofii RAN, Moscow, 1993, p. 81. (*)
[14b] *ibid.*, pp. 107-116.
[15] **Petrenko, V. & Mitina, O.**, "Psychosemantic studies of political mentality in Russia: 1991-1993", *Obshestvennyje nauki i sovremennost*, **6**, 1994, pp. 42-54. (*)
[16] **Rosenthal, B.**, "The occult in modern Russian and Soviet culture: an historical perspective", *Theosophical History*, **8**, 1993, p. 259.
[17a] **Sheregi, F.**, "Perspectives of development of Russian ethnic self-consciousness", *Vestnik RAN*, **1**, 1995, p. 50. (*)
[17b] *ibid.*, p. 51.
[18] **Shikhirev, P.**, "Dynamics of socio-psychological state of Russian society", *Psikhologicheskij Zhurnal*, **3**, 1993, p. 143. (*)
[19] **Sorokin, P.**, *A Long Way: Autobiographical Novel*, Soyuz Zhurnalistov/ MP "Shyvas", Syktyvkar, 1991. (*)
[20] **Tart, Ch.T.** (ed), *Altered states of consciousness*, Anchor Books, Garden City, 1972.
[21] **Toffler, A.**, *Powershift: Knowledge, wealth, and violence at the edge of the 21st century*, NY-Toronto-London-Sydney-Auckland, 1991.
[22] **Tutorskaja, S.**, "In response to the insanity of the world", *Izvestia*, 4, 22.12.1994. (*)
[23] **Wolman, B. & Ullman, M.** (eds.), *Handbook of states of consciousness*, Van Nostrand Reinhold, NY, 1986, pp. 286-287.
[24] **Zdravomyslov, A.**, "Modernization of Russia from the point of view of conflict sociology", *Vestnik RAN*, **4**, 1994, p. 301. (*)

(*) sources marked by asterisk are in Russian.

DIEDERIK AERTS

SOCIAL SPACE: FROM FREEDOM TO FREEDOM OF MOVEMENT

The first man who, after he had enclosed a piece of land, had the idea of saying "this is mine", and encountered people who were simple enough to believe him, was the real founder of the political society. How much crime, war and murder, how much misery and terror the human race would have been saved if someone had pulled out the posts or filled in the ditch and called out to his fellow humans, "guard against listening to that cheat; you will be lost if you forget that the fruits are for everyone, and that the Earth belongs to no one."

Jean-Jacques Rousseau in 'Du Contrat Social'

Since ancient times, everyone has longed for an idyll, for that garden where nightingales sing, that harmonious domain where the world is not strange to man and man not strange to other men, but where, on the contrary, the world and all its people are created from one and the same material and where the fire that burns in the heavens is the same as burns in the human soul. There, everyone is a note in a brilliant fugue by Bach, and anyone who does not want to be, will remain just a black spot, a superfluous spot, robbed of purpose, and which one can pick up and squash between your nails like a flea.

Milan Kundera, 'The Book of Laughter and Forgetting'

The boy ran through the desert for forty days until he came to the splendid castle on top of a mountain. There lived the wise man the boy was seeking. He walked into a hall full of tremendous hustle and bustle: merchants were rushing in and out, groups of people stood talking in every corner, a small orchestra was playing charming melodies, and a plentiful table had been laid, overflowing with all the most delicious dishes from that part of the world. The wise man was talking to everyone and the boy had to wait for two hours before it was his turn. The wise man listened attentively to the reason for his coming, but told him that unfortunately he did not have time at that moment to explain the secret of happiness to him. He suggested he go for a two-hour walk around the castle and to come back afterwards. "I would however like to ask you something," said the wise man in conclusion, as he handed the boy a teaspoon from which hung two drops of oil. "I would like to ask you, while you are walking, to hold this spoon so that the drops of oil do not drip off." The boy began to walk up and down the stairs of the palace with his eyes fixed rigidly on the spoon. After two hours he returned to the wise man. "And," asked the latter, "did you see the Persian carpets in the dining room? And the garden that took ten years of the head gardener's life to complete? And the superb parchment scrolls in my library?" The boy shamefully admitted that

© 1999 *Kluwer Academic Publishers. Printed in The Netherlands.*

he had seen nothing. His only concern had been not to spill the drops of oil the wise man had entrusted him with. "Well go back and acquaint yourself with the wonders of my world," said the wise man. "You cannot trust a man unless you know his house." The boy started off, walking around the palace all over again, but this time he looked at all the works of art on the ceilings and walls. He saw the gardens, the surrounding mountains, the beauty of the flowers and the sensitive arrangement of the works of art. When he returned to the wise man he gave a detailed description of what he had seen. "But where are the drops of oil I entrusted you with?" asked the wise man. The boy looked at the spoon and saw that he had spilt them. "This is the only advice I can give you at this time," said the wise man. "The secret of happiness lies in looking at all the wonders of the world without ever forgetting the two drops of oil on your spoon."

Paulo Coelho in 'The Alchemist'

A week ago the teacher and his students had met each other on the lawn of the garden of Academe. On this occasion they had agreed to meet on the Grand'Place in Brussels. They were looking for a table on the first floor of the café called 'Le roi d'Espagne', with a view over the splendid square. While the waiter was taking their orders a conversation developed.

"After our last discussion I read a lot of newspapers from the second half of the twentieth century," Ernest said, "and it was remarkable how universal the problem of unemployment was. Creating employment was a priority for every government. Many European, American and Japanese companies were obliged to employ fewer and fewer workers in order to remain competitive on a world scale."

"Surely that's strange," said Sonia, "why did they have to dismiss workers to remain competitive?"

"It was a complex situation, one we can hardly imagine these days," said Jan. "The governments of the Western countries, mostly social democratic ones, had committed themselves to developing and maintaining a social system which derived most of its income from tax on work. This tax was paid partly by the employer and partly by the employee. In the second half of the twentieth century the system had grown so much that the employer had to pay almost twice as much as the employee actually received. So, in concrete terms, for every 100 Euro of a worker's net earnings, another 100 Euro went to the state, so the employer had to pay 200 Euro."

"Wasn't that the same in all parts of the world, then?" asked Ernest, "and was it for competitive reasons that people had to be dismissed?"

"Yes, it was," continued Jonito, "since the less developed countries, which had a far less sophisticated social system, could provide much cheaper workers. This brutal fact set an initial change in motion: big multinational companies, which had the resources, moved to the less developed countries, where they could manufacture the same products much cheaper."

"There was also a second reason," said Jonito. "The increasing role of technology in production made it possible to have machines do a large proportion of the work formerly done by people. This also saved money, because a machine only has to be maintained, not paid for its work. That's why the big multinational companies systematically switched to more machines and fewer workers."

"And in addition to that we should also mention here the mechanism of the repeated and almost unending economic recessions," remarked Ernest.

"Yes," said Jan, "the fact that a lot of people became unemployed, and thereby ended up in the social security system, meant that the state needed more and more of the money they wanted to use as a safety net for them, and so it became more difficult to make savings. On the other hand, savings had to be made if the system were not to collapse completely, and this meant that the employees who were not unemployed were paid less and less, which reduced their purchasing power, so that the market did not function so well, creating even greater problems for highly competitive companies. And so the cycle of recession was complete."

"Now I understand why workers had to be dismissed in order to remain competitive," said Sonia, "but I have a second question, which is, why for heaven's sake, was it necessary to remain competitive?"

"That was a situation that was not really questioned at the time," replied Jan.

"Yes, I understand what you mean, Sonia," said Jonito, "why didn't the states protect their own companies from this antisocial competition from low-wage countries?"

"I think you have all stated the problem very well," said the teacher, "and it was indeed a distressing situation, a downward spiral which was hard to get out of. And your second remark was very interesting, Sonia, we shall have to think about that one."

"Last time you told us," said Sonia, "that the transition from an observational to a participatory paradigm finally brought a solution to this situation. You also pointed out the importance of the change in education in this respect [1]. But I can no longer see so clearly how this change of paradigm played a part in solving this problem."

"Stated in the clearest form," said the teacher, "unemployment was a problem of participation. The society of the time did not succeed in allowing people to participate in the life of society. As the participatory paradigm became stronger and more knowledge and experience was acquired about the way participation could be achieved, it also became easier to solve this problem. But even so, it was a very complex one. Perhaps we should reflect on that today, the subtle way the change from an observational society to a participatory one was the basis of the solution of the crisis situation at the end of the twentieth century." The teacher paused for a moment and the students nodded in assent and interest. In the meantime the drinks had appeared and everyone had taken a glass. The teacher continued. "In order to make the problem clear I must first tell you about the essential difficulty of 'maintaining a social entity'. Since we now live in a society that is organised on much more participatory lines, we are probably no longer aware that 'maintaining participatory collaboration' is no trivial matter. These conditions had to be fulfilled in order to develop into a participatory society. Experience, knowledge and understanding in this area only came about in small steps."

"Does that mean the desire for co-operation is not the main issue?" asked Ernest.

"It may seem like that from the comfortable position we now find ourselves in," said the teacher, "but in fact it is definitely not the case. A subtle balance of forces and dynamic changes is necessary to maintain participatory collaboration. The nature of these forces and the laws of these dynamics have only very gradually become clear. Due to the great confusion that prevailed regarding this issue it was only at the very end of the previous century, when the spontaneous movement from an observational society to a participatory society was already up and running, that actors and thinkers started to guide this evolution in a more or less systematic way. To this end, all sorts of measures were taken to solve the crisis, most of which were aimed at the short term and thereby sometimes even hindered the move from an observational to a participatory society. If we want to see clearly what happened at the time and how we arrived at the situation we now have, we have to list a number of aspects."

"Surely there has been collaboration as long as humans have existed?" asked Ernest.

"Yes, there has," replied the teacher, "but 'collaboration' did not always mean 'collaboration in a participatory way', and that is what it's all about. There are various ways of maintaining collaboration, and the participatory is only one of them. There is also the hierarchical way, for example. We do not want to call purely hierarchical collaboration participatory."

"But surely there is collaboration in hierarchical organisation too?"

commented Ernest.

"Quite right," replied the teacher, "a hierarchical way of organising things allows collaboration, as has often enough been demonstrated throughout history. But that way of collaborating is very specific: in a purely hierarchical society all explicit decisions in connection with possible actions will come from a central point, and the majority of the individuals taking part only carry out these decisions. Although it is a way of co-operating, we shall not call it participatory. By participatory we mean that every individual takes part in co-operation with his or her whole being (as far as possible): as co-actor, co-decision-maker, co-executor, co-observer, co-planner and co-thinker."

"It's true, that situation is certainly not achieved in hierarchical collaboration," said Jonito.

"Humankind has come a long way in achieving the situation as it now is," continued the teacher, "and perhaps we should also discuss the various structures involved in this path if we want to have a clear view of the dynamics of participation."

"That sounds like a very good idea," said Ernest.

"Traditional societies have often had participatory forms of collaboration," said the teacher, "and that is why they have often been idealised."

"What do you mean by a traditional society?" asked Jan.

"Well," continued the teacher, "for the sake of simplicity, since history is naturally much more complex than the outline we are proposing here, let us call a traditional society that sort of society which is organised mainly on the basis of a fixed social structure, sometimes hierarchical, but not necessarily so. In a traditional society every individual had his fixed 'place' within the forms of co-operation that existed there. These forms of co-operation were chiefly determined by tradition—that is where the name comes from—and were kept stable and bearable by a covering of rituals and customs that were sustained by this tradition."

"Were most old civilisations in that way traditional societies?" asked Jonito.

"You could put it like that," said the teacher, "although I would like to say once more that we should beware of employing too general an outline. The break with this traditional way of living together started slowly and reached a high tempo during the rise of modernity in Western Europe in the 16th and 17th centuries. In the 20th century this movement away from traditional communal living drew the whole of mankind along with it. Traditional societies are characterised by the specific way they succeed in stabilising forms of collaboration. In the classification I want to employ, traditional societies are often participatory, but generally static. The social entities they stabilise are not particularly mobile within social space, but they are stable and often participatory."

"What do you mean by that?" asked Ernest.

"In the analysis I would like to propose, the concept of 'social space' is an important one," said the teacher.

"But what do you mean by 'social space'?" asked Jonito.

"We shall talk about that later," said the teacher, "now I think it's time we went for something to eat.

MOVEMENT IN SOCIAL SPACE.

The teacher and his pupils walked through Brussels looking for a restaurant. They sauntered across the Grand'Place and admired the uniformity of the Town Hall, the Breadhouse and the surrounding houses, all of them dedicated to particular medieval guilds. They then walked towards the Beenhouwersstraat and looked for a quiet table in one of the many restaurants. The conversation then resumed.

"I must explain to you what I mean by 'social space'," the teacher started. "Social space is the collection of potentially differing social patterns that individuals and groups within a social organisation can adopt. I want to use the word 'space' deliberately here, because the physical space in which we move with our bodies is in fact also the collection of all the potential places where objects and collections of objects can be in relation to each other."

"Is that really a useful analogy?" asked Jan.

"Certainly," answered the teacher, "it is possible to examine the nature of physical space in a very fundamental way [2], and then one comes to the conclusion that on the basis of this analysis it is correct to introduce the concept of 'social space'. If a certain individual or group of individuals changes their social pattern in a social context, we shall interpret that as if they are moving in social space. You will notice that this description puts in a new light the changes in the forms of human society throughout human history, and explains a lot of apparently contradictory aspects of it."

"So is social space the collection of all possible forms of communal living?" asked Jonito.

"That's right," replied the teacher, "you could consider social space in its totality in that way. In the rest of our talk it will become clearer in what way we can understand certain phenomena in this new context, and that will also make it clearer what 'social space' is."

"Can you give us an example?" asked Ernest.

"Well," said the teacher, "let us consider the way traditional societies, which we have already discussed, are present in social space. We could say that these traditional societies were mainly static in social space. The social pattern and individual experiences in such a traditional society, as well as the network of patterns in a group in this traditional society, are very much determined by tradition and customs and subject to little movement. In a traditional society every individual and every group has its clearly defined place and hardly moves in social space. We should note, however, that locally, in the proximity of this place, the individual has a great deal of individual freedom of action and growth. This form of society is stable and in a particular way has averted many possibilities of internal conflict. In fact, since the place is allotted by tradition and is thereafter no longer questioned, the freedom an individual can experience in one such clearly defined place is never under threat, nor is it a threat to the freedom of the others. Social space is large and proportionally well filled, but there is little movement within this social space: it is a social space filled with static social patterns. Since individuals are allotted other social patterns as they get older, there is actually a systematic movement, but it is controlled by tradition and customs. So one might generally say there is markedly little 'free' movement. If we look for an analogy with physical space, we might compare the fixed places social patterns have in traditional societies with the fixed places trees—physical patterns—have in physical space. Once a tree has been planted, its position in physical space remains fixed throughout its existence. The changes in the tree consist only of 'growing' in relation to the other trees in the wood. In this analogy the rules of tradition correspond to the ecological laws of the wood. These traditional cultures are often very hospitable to foreign visitors, because they have provided a 'fixed' place for the patterns of such visitors in their social space. This visitor can move very freely within this position in social space, and participate in society while still remaining himself."

"The example of the tree is interesting," said Jan, "but I would like to say that trees and plants do exist in a spatially static relationship without craving each other's place in the field. But there clearly is competition for the space in the third dimension, and a tree even has the capacity to spread germicidal substances so that it is difficult for any competitors in its immediate surroundings."

"That is true," said the teacher, "but it is only an example, intended chiefly to illustrate the difference between animals and plants in their presence in physical space."

"I see where you are leading," said Jan, "the social patterns of individuals in modern forms of society do of course move within social space."

"Yes, that's right," said the teacher, "and as we continue our story we shall see that the fact that the social patterns in a modern society do move is the cause of a lot of extra difficulties. It is not surprising when one reflects on it. Working together in a society of moving social patterns is naturally much more difficult than when they are static. Trees in a wood have to 'concern' themselves with little more than the ecological balance of the whole. For example, they cannot take each other's place, since changing position is not possible for them, and so their 'own' place is never under threat. This is why there is little envy (the desire to be or become like another), since this possibility barely exists in traditional societies" [8].

"So it was the possibility of 'moving' in social space that gave a free hand to negative forces like envy?" asked Sonia.

"That's right," said the teacher, "though envy existed before that. Remember that in the Old Testament it was branded as one of the greatest sins. But in a traditional society social space was conceived in such a way that the destructive aspects of envy could be kept more or less under control. One of the main elements in this control was the static aspect of social space in the traditional forms of society. However, it is clear that in the traditional societies from which we are still able to collect data, the individual's attempt to move his or her social patterns within social space had already started. For that reason envy was already recognised as a dangerous force which would have a disintegrating effect on the whole society if its negative aspects could not be checked. The actual birth of envy, which took place thousands of years ago, was the first sign of the start of movement in social space, at the same time as the awakening of social awareness among social mammals, and more particularly human consciousness among hominids."

"But what was there before that?" asked Ernest.

"We still do not have a clear idea of that," replied the teacher. "It could be one of the more important objects of study in social biology. If we look at the social insects, the ants, termites and bees, it turns out that the social structure is maintained chiefly by the instinctive behaviour of the individual insects [13]. More primitive social mammals, such as wolves, also act within the group in a way that is primarily determined by instinct [19]. Instinctive social behaviour was probably the predecessor of the social behaviour that was co-ordinated by customs, rituals and traditions. All these more primitive social structures are static and do not move in social space."

"So movement in social space is the main element distinguishing the new, modern behaviour from the previous forms of social organisation?" asked Jonito.

"That's right," said the teacher. "When the impulse towards movement was already present, but was still seen as a great danger, threatening the disintegration of the existing static society, most thinkers actually rejected this movement. Aristotle, free of complexes, defended the image of a society ruled by a 'natural hierarchy'. Everyone has his place in a naturally organised cosmos, and men and women have their own place, the man's place being 'higher' than that of the woman. Some people are by their nature more suited to ruling and others to serving [24, 25]. In 'The Republic' Plato also describes why a natural hierarchical arrangement is necessary to organise society in balance and harmony. He pressed for a form of state in which society was ruled by a small elite" [26].

"The way you describe it to us now sounds as if it was all simpler and functioned much better in those static, traditional societies," commented Sonia.

"That was partly true," replied the teacher, "you should not forget that the static, traditional societies had themselves slowly developed out of earlier forms of communal living that were based more on instinctive social behaviour. This made them very well-balanced and stable and they had absorbed rituals and traditional mechanisms that were able to ward off great disasters. To give you an example, there was the mechanism of the 'scapegoat'. In the twentieth century this scapegoat mechanism has been described at great length by the French philosopher René Girard [11]. According to Girard, envy could also suddenly flare up in a traditional society. Let's imagine the situation where several people want to possess the same object. It could be food or any other valuable object or even a woman. This dispute may then develop further and envy may then infect the whole group. When a peak of disorder is reached all the aggression and animosity is often concentrated on one individual in the group. This one person attracts, as it were, every calamity and is eliminated by the others. This sacrifice means the aggression departs all at once: killing the scapegoat brings about catharsis for the whole group. Harmony is restored because the person identified with the calamity has been removed. According to Girard, this scapegoat mechanism was fundamental to traditional societies, and the scapegoat, whose sacrifice has returned harmony and union, was later often deified. Since the rituals and customs were the cement that held traditional societies together, the scapegoat became an exemplary figure in the ritual. The process as a whole will on the one hand serve as a model for the ritual it wishes to repeat (on the basis of the principle that one must always repeat what the victim, as long as it was salutary, did or underwent) and on the other hand as a counter-model for the prohibitions, on the basis of the principle that one should never repeat what this same victim did, insofar as it was malicious. In this way, disaster and the danger of disintegration of the entire society is stored in the ritual memory of the traditional society on the basis of the sacrifice of scapegoats."

"I have studied Girard," said Ernest. "I was indeed greatly surprised to learn that this mechanism existed in many primitive societies and was even considered by Girard to be inseparably linked to the origin of the culture. I still remember that scapegoats were usually people who were special in some way, so that they were distinct from normal individuals."

"That's right," said Jonito, "throughout the world, mythology is teeming with cripples, people with one eye or one arm, twins and others with a special characteristic. Girard claims they were often the object of the scapegoat mechanism."

"Heroes and gods also often played a part in a mythology that can be traced back to an original scapegoat incident," continued the teacher. "Mythology often tried to cover up the original violence of the murder of the scapegoat so that in later stories these figures became gods or heroes. When the change to a dynamic form of society was set fully in motion, these old mechanisms turned out not to function any longer. Changes like that are always dramatic events. One of the first philosophers who reported on this at length was Thomas Hobbes in 17th-century England."

"Hobbes? Was that the man who proposed the great Leviathan?" asked Sonia.

"That's right," replied the teacher, "Leviathan was Hobbes' greatest work and it was there that he focused on the failing mechanisms of what he called a 'natural' society [27]. According to Hobbes, 17th-century English society had degenerated into a 'condition of war of everyone against everyone', and this war was kept going by 'envy'. It is also interesting to note that Hobbes' analysis is based on an analogy with the laws of mechanics formulated not very long before by Galilei. For example, Hobbes said that something that lies still will always remain lying still, unless something else sets it in motion. This is a truth that no one doubts. But the idea that something in movement remains forever in movement unless something else brings it to a standstill, is not so easily accepted. Galilei had in fact formulated the law that claims that an object in uniform rectilinear motion will always remain in motion if one does not bring any force to bear on it to change its state of movement. Drawing inspiration from this law, Hobbes stated that people's desires are always in motion and only come to a standstill if compelled to do so by a force. Hobbes is here referring to envy. Hobbes proposes a solution that 'regulates' this 'uncontrollable' movement, and that is the great Leviathan, a powerful form of state that issues laws and regulations that are able to check this pernicious movement caused by envy."

"It is odd that Hobbes used the concept of movement," said Jan.

"Yes, it is," said the teacher, "but although he was certainly not thinking in terms of movement in social space, the way I am trying to clarify these events for you, the inspiration he drew from Galilei's discoveries did point in the same direction. One might say that the most important later social philosophers have again and again struggled with this problem that Hobbes brought up in a most intriguing way. They were always looking for solutions that would give a structure to these uncontrolled movements in social space."

"Without being aware that they were dealing with a new state of 'movement'?" asked Jan.

"One could put it like that," said the teacher, "although on the other hand it is also clear that most philosophers agree that one cannot return to a traditional static situation. A solution has to be found for the new situation. This solution of course exists: it comes down to finding a situation in which the dynamic situation is stabilised: in which individuals learn to move their social patterns together in social space without posing a threat to each other's movement. However, it appears that there were plenty of hidden snags to impede the establishment of any such stable dynamic situation. A lot of problems were created by the rising 'concept of equality' and the misunderstandings this way of thinking brought about."

THE CONCEPT OF EQUALITY: MISTAKES AND DIVERSIONS

The students were astonished by the teacher's last statement. 'The rising concept of equality' was a phrase that appealed to their imagination, though they suspected that the teacher used it primarily to draw their attention. The teacher felt the tension and paused while taking a deep breath.

"The rising concept of equality," said Jan, "that was surely a very good thing."

"One cannot put it so simply," said the teacher, "the concept of equality arose as a reaction to the movement that had started from the static traditional society to the dynamic modern one. We must realise that in a traditional society there was no 'inequality' because the act of 'comparison' was largely absent. Let's have another look at the situation of a wood and its trees. The trees are not unequal because they are in different positions. The problem of the equality or inequality of the trees does not arise because the trees cannot move into each others' positions and so the differences between the positions cannot therefore be compared by the trees themselves. The mathematician and physicist Henri Poincaré, who lived in the late 19th and early 20th century, and was also involved in the foundations of relativity theory, analysed this situation very thoroughly on the basis of a highly formal approach. He came to the conclusion that knowledge of the reality of space is linked to the possibility of moving in that space [28]. So we are now able to note the differences in the position of the trees because we look at the wood from a dynamic situation and we are able to move about in the physical space in which these trees take up various positions."

"I see what you mean," said Jonito, "since the traditional form of society is primarily static, no comparison is possible in connection with differences that only appear as a consequence of movement in a dynamic form of society. In the static traditional society there was hardly any knowledge with regard to the existence and the nature of social space."

"That is correct," said the teacher, "and so it was only when the static pattern began to move, and at the same time social space was created, that comparisons became possible in connection with differences that arose as a consequence of this movement. The result of the new situation was that one and the same place in social space could be desired by several individuals. And it was therefore also meaningful to compare the various positions since it was in principle possible to force someone out of their social position and take up that position oneself. The battle for the various social positions was able to commence. It was mainly due to this that the old force of envy was given a great new field of action. The concept of equality was also a consequence of this new state of movement."

"I still have the feeling this concept of equality was very beneficial," said Jan.

"And you are right," said the teacher, "the concept of equality was borne along by a beneficial force, but can easily lead to a great many mistakes and derailments. If the concept of equality states that we are all essentially 'equal', then it is arguing, among other things, for a homogeneity and isotropy of social space at the deepest level, and that is a very good requirement, which we in fact also intuitively sense to be so."

"Yes," said Jonito, "I see what you mean. Physical space, which you have used as an example several times, is also homogeneous and isotropic. Every position is potentially equal for every object that wants to occupy that position."

"That is right," said the teacher, "the analogy brings great clarification to this point. Every position in physical space is equal for every object that occupies that position, and when an object moves in space it leaves its position intact and with the same potential for a future occupant. If the aspiration of the concept of equality is a homogeneity and isotropy of social space, this is an aspiration to be applauded, but there are often a great many mistakes made in the process. Equality should not mean that we all want to occupy the same position in social space. If that is the case, the only consequence of the concept of equality is a shrinkage in social space. In the most extreme case social space implodes to one single position or a very limited number of available positions."

"But homogeneity and isotropy of social space is not enough to implement the idea of equality," said Jan. "I find the meritocratic principle to be a more complete basis for the concept of equality, whereby transition within the social structures takes place purely on the basis of merit, regardless of the identity of the individual. The fact that the transition of the individual in social space is independent of his identity does not so much say anything about the metrics of social space. The transition from one position to another in this space is dependent solely—in the best of cases—on the two positions themselves. In a less favourable case the transition would be determined by the history of the entity, right back to the situation at its start. Determination in the cradle is not non-existent."

"You are absolutely right," answered the teacher, "Homogeneity and isotropy are only possible conditions. If we again think of the analogy of physical space, the movement of a material object in that space is subject to yet other symmetries. It would make an interesting study to disentangle the various symmetries that characterise the concept of equality in physical space."

"I can easily imagine the possible mistakes in connection with the concept of equality," said Ernest. "After all, in his work '*Het rijk van de schaarste*' (*The Realm of Scarcity*), Hans Achterhuis made a penetrating analysis of the ways the concept of equality has resulted in scarcity [29]."

"I know Achterhuis' work," said the teacher, "and it certainly inspired me when it came to the image of social space I want to bring out here. Achterhuis thoroughly analysed the various incorrect ways of interpreting equality, and he does indeed demonstrate how they always lead to 'scarcity'. Scarcity, as Achterhuis treats it, corresponds to a diminishing of social space. The situation of a dynamic form of society is even more complex than an analysis of the concept of equality and scarcity can demonstrate. What it is about is how humankind learns to move in this social space in a harmonious manner. The example of physical space and how objects move within it is able to show us the wealth of possibilities. It's true that objects can move in many different ways in relation to each other. But it is clear that an object cannot take up the position of another object while that object is still there."

"Objects can collide and one can force another from its position," remarked Sonia.

"That's true," said the teacher, "but if we consider the movement of the bodies of people and animals in physical space, in general very little serious damage is caused by such collisions. Any body that tries to oust another body from its position by violence and against the will of that other body, is in general acting in a way that is experienced as not ethical."

"And then there were the traffic accidents in the twentieth century," commented Sonia. "We have already talked about unemployment in the late twentieth century, but there was an even greater scandal at the time. I recently read that an average of an enormous amount of people per year—I do not remember exactly how much—most of them young adults and children, were the victims of traffic accidents. Those were accidents that involved only movement in physical space."

"I am glad you mention that situation," said the teacher, "because we can see there how a very old stability can be upset by new developments. The fact that our bodies, and those of other moving living beings—which are in fact sets of millions of co-operating cells—are the way they are, has to do, among other things, with the fact that they have to withstand collisions with other moving bodies and objects. One of the causes of this is what is often called Darwinian evolution. When man began to use tools a new element appeared. Tools and weapons made of stone, and later of metal, were a threat to the old balance. It is possible to shatter someone's skull with an axe and thereby literally oust him from his position by killing him. The development of traffic in the twentieth century, and particularly the use of the car, is a technological continuation of this evolution. Man started moving about in cars in a way that was not in balance with the construction of his body. This has resulted in the enormous and incomprehensible harm done by this evolution. There was hardly any resistance to the rise of the car and dangerous traffic, since this change took place in an area of reality that had already been stabilised hundreds of thousands of years previously and where there was therefore no longer a 'struggle' for stabilisation."

"So is mankind's task to learn to move in a harmonious way in social space?" asked Sonia.

"Yes, it is," replied the teacher, "and a whole lot of new aspects are involved, and again the analogy with what once took place in physical space is able to throw light on the subject. It is clear that if, as a group, we move with our bodies in physical space in a peaceful and pleasant way, there is no need at all to move into each other's positions. If a group can sit round a table, as we are now, its members will have no need to deprive each other of their places. In fact in this case we all expressly wish to sit in different physical positions, since it is only possible to have a conversation like the one we are having now because of the explicit differences between the places. The difference in physical position and the lack of inclination to take each other's place is one of the conditions for productive co-operation. The example we are considering now, of ourselves sitting at the table, is a static example. And it's true that in general positions are not changed, although it is possible in principle. The situation at a reception, where everybody walks round and where new conversation groups are constantly forming is an example of a dynamic situation. There are a great many examples of dynamic situations in which various people together peacefully move their bodies in physical space: sport, dance, theatre, walking, etc. What makes the being or working together interesting and productive is the way the various places where the individuals move are related to each other in the eyes of the group."

"Yes," said Jan, "it is no coincidence that it is only in fights and situations of human violence that a real attempt is made to capture other people's positions."

"Yes, that's true," said the teacher, "it is no coincidence that the customs that harmonise the movement of people's bodies in physical space are violated when two individuals attack each other violently."

"But how do you see a thing like that working in social space?" asked Ernest.

"Well, let's think about that for a moment," said the teacher. "We have to clearly identify what 'equality' means in social space. It should be the condition for the construction of a homogeneous and isotropic social space, in which any social position is in principle accessible to everyone. This means that all social positions also offer everyone the same fundamental potential, which is what makes them into positions in social space, if there is movement towards them, and in which harmonious movement is possible for everyone. There are two essential conditions to be able to achieve this sort of equality and this sort of movement. The first condition has to do with the amount of space: social space has to be large enough to provide a comfortable place for all patterns of social interaction in which an individual or a group of individuals wants to move. The second condition has to do with stability: social space must possess those qualities that make stability and co-operation possible and which also guarantee stability of movement. We shall see in the course of our conversation that neither of the conditions is fulfilled in a commonplace way, and that their accomplishment has far-reaching consequences for the nature of society."

"That will be fascinating," said Jonito. By this time the teacher and the students had finished their meal and the waiter came with the bill. They then left the restaurant and walked through the King's Arcade where a few musicians were playing sonatas by Telemann on the recorder and guitar.

STABILITY OF SOCIAL PATTERNS OF MOVEMENT

They decided to have some dessert and a drink in the Mort Subite, a traditional Brussels café, where it is still possible to have a pleasant conversation without being bombarded by Japanese decibels. They ordered a few beers and the conversation resumed.

"What are the conditions under which a social entity is stable and can move in a stable manner within social space?" asked the teacher. "That's an issue we should explore now."

"Or how can people work together?" reacted Ernest.

"Quite," said the teacher. "That's another way of phrasing the question. And the conditions for fruitful and stable co-operation are certainly not trivial. In fact, different aspects of this basic issue, the stability of a social entity or how forms of co-operation are structured, have been studied by various academic disciplines with an interest in the subject, such as sociology, economics and psychology. A typical situation in connection with this issue has even been highlighted: the prisoners' dilemma."

"What is the prisoners' dilemma?" asked Jan.

"Imagine there are two prisoners," replied the teacher, "who have committed a crime together and have been arrested. They are interrogated in separate cells so that they cannot agree on a story. They are told that the first one to confess will be set free as a reward for helping the police with their enquiries. The one who does not confess will get a ten-year prison sentence. If neither of them confesses, there is enough circumstantial evidence to convict them anyway and they will both be sent to prison for one year. If both confess, they will each be sentenced to five years in prison."

"What a terrible situation," said Sonia.

"A fascinating situation," said Jonito. "The best solution for both of them would be for neither of them to confess."

"Exactly," said the teacher, "but that solution is not stable, in the sense that there is still the risk that the other one will confess, leaving his accomplice with a 10-year prison sentence."

"It's not really a very realistic example," remarked Sonia. "The world is a bit more complex than the rather artificial situation that these two criminals find themselves in."

"You are quite right," said the teacher. "Personally I do not really like the prisoners' dilemma and the way it is constantly used as a example. Nevertheless we can recognise in it the basic issue of the stability of social entities. If I may, I would like to look at this basic issue separately from the example of the prisoners' dilemma, and describe it in more general terms. Assume we are considering a general situation involving co-operation between various individuals. If there is co-operation, this means that the actions and decisions of one of the individuals will be dependent on the actions and decisions of the other individuals. If this is not the case, then all we would have is a collection of individuals who are not working together, but are simply together, like separate, fragmented entities in social space. This situation of non-co-operation is a limit situation of the co-operation situation, that is to say the limit situation where the interactions between the individuals are negligible. The nature of the interactions will of course determine the nature of the possible dynamic configurations to a large extent. Let us just consider the situation of a deterministic interaction. That is the situation in which objects, and hence also our bodies, find themselves in physical space. If we mention this, it may become increasingly apparent why the comparison with physical space, a comparison that we have made several times, is relevant. Physical objects in physical space have developed all kinds of dynamic and participatory stable interaction patterns, which are all, however, controlled within a deterministic framework. Deterministic means that the behaviour of objects in this physical space is 'determined' by the states of these objects and the interactions between them. The laws that deduce this behaviour from states and interactions are the laws of classical mechanics developed by Isaac Newton [30]."

"But this determinism is only a feature of classical mechanics and not quantum mechanics," remarked Jonito.

"That is not quite true," said the teacher. "Quantum mechanics demonstrates that the behaviour of quantum entities, photons, electrons, protons, neutrons, atoms and molecules, is no longer determined only by the states and interactions of the entities in question. There is an essential element here, which emerges as a consequence of the way these quantum entities can locate themselves in physical space. Quantum entities seem to have a fundamentally new aspect, which is related to the way in which they can place themselves in a particular macroscopic structure, and this aspect is mistakenly associated with indeterminism [31]. It often happens, particularly when quantum entities move from one macroscopic structure to another (photons moving from one atom to another, for example), that these quantum entities find themselves in a non-local state, and are not located within physical space [36]."

"You mentioned this non-local behaviour of quantum entities earlier," said Jonito. "Do you think there is an analogous non-local behaviour of entities in social space?"

"Wait a second," said the teacher, "that is a bit of a leap. But it is a question I definitely do not intend to avoid. There are analogies that point to a very profound connection, but first I have to tell you about how a social entity can stabilise itself. Once we have explored that in depth, we shall come back to the question of quantum entities. Incidentally, I should mention at this point that it still has not been proved that social behaviour is non-deterministic. That is an intuition that we may personally feel because we think that we have a 'free will' and can make 'free choices', which are 'not determined' by the states and interactions that are found in the world around us. But the way in which this 'free will' is connected with determinism or indeterminism is another issue that we need to talk about sometime." The teacher paused and took a sip of his beer, then continued: "The stability of a collection of moving material objects in physical space is given by the solutions to Newton's equations, and we know several good examples of these."

"Which ones do you mean?" asked Ernest.

"Think about the movement of the Earth around the sun," said the teacher. "Incidentally, the question as to how this structure of movement could be stable was one of the greatest theoretical problems ever solved by mankind. As a link between forms of co-operation in social space and the determined way in which material objects move together in physical space, it consists in co-operation between lower social animals based on instinctive behaviour, which we have already talked about. The extent to which instincts determine this behaviour is also open."

"But if we do not know whether the same mechanisms are at work," volunteered Jan, "can we really compare the different types of stability conditions?"

"We have to proceed cautiously," said the teacher, "but you will see that the mechanism of stability does not fundamentally depend on the type of interaction. So far we have talked about the negative power of 'envy' in a far from specific way because of the possibility of jeopardising forms of co-operation. That is an extreme simplification of the real situation. One might also be led to suspect that this negative power is one that is only present in a particular individual, namely the individual who is envious. But this is only a tiny part of the story. Let us therefore try to describe the disintegrating and destructive force for co-operation in social space in a more general way, and then look at how parts of it are given other names depending on the situation in which they occur. Let's assume that we are considering a group of individuals who want to carry out a specific project as a joint venture. The 'envy' that we mentioned above is just one aspect of the various ways in which these disintegrating forces can materialise: it is the situation where one of the individuals covets the position that another individual occupies in the joint venture. This is one of the simpler manifestations of the destructive possibilities. We came across another example of the possibilities of destructive forces in the prisoners' dilemma. The main emphasis here is on the fact that often the one who takes the first step towards destroying a social entity can immediately benefit greatly from it personally. Let me give you an example that illustrates the same dynamics. Suppose a group of shopkeepers agree between themselves that they will only open for business three days a week, since they have all realised that they can sell everything on those three days that they would normally sell in a whole week. This mutual agreement will give them more free time to improve the quality of their shop and of their own life, with the result that they all win if they stick to the agreement. It is clear, though, that if one of them does not stand by the agreement, he will be able to make a lot of profit initially. That is why the situation of this kind of agreement is fundamentally unstable."

"Yes, I understand that," said Jonito, "and it's clearly a pure form of instability. If one of the shopkeepers decides to go back to opening his shop all week, some of the others will quickly follow his example because they feel that the contract has been breached and they should not have to be the victim of such a breach. This triggers, as it were, an avalanche, with the result that even if there were still a few shopkeepers who wanted to stick to the agreement, they would find themselves forced to give in and open their shop all week again, because otherwise they would be marginalised and would really be at risk."

"That's right," said the teacher. "That's a good description of the instability of the shopkeepers' situation. The one shopkeeper who breaches the contract does in fact trigger an avalanche and no-one seems to be able to escape it. You no doubt feel that this example of the shopkeepers has brought us much closer to a general characteristic of the possible destructive forces. Perhaps you are also better able to understand why I emphasised the word 'stability' with respect to a social entity. And I would like to consider again the useful analogy with the situation of interacting material objects in physical space. Classical mechanics gives a definition of stability and instability that applies perfectly in social space. The definition is as follows: suppose we are considering a number of interacting objects in physical space that are in a certain state. It is possible that this collection of objects is subject to a 'disturbance' so that they are shifted slightly out of this state. As a result of this disturbance and change in state, new forces may spontaneously arise, which influence the collection of objects. We shall assume that the original state these objects were in is 'stable' if the natural forces that arise as a result of disturbing this state act in such a way that the disturbed state will shift back to the original state. If, however, the forces that arise as a result of the disturbance tend to shift this disturbed state even further away from the original state, then we call this original state unstable."

"Just a minute," said Sonia. "Can you give us an example to clarify that?"

"There is a very simple example that will illustrate this mechanism," said the teacher, "and that's the example of the pendulum. If we consider a pendulum in its equilibrium state —in other words hanging straight downwards—this is then a stable state for the pendulum. Now suppose that we move the pendulum out of its equilibrium state; the forces that arise are such that they will bring the disturbed state back to this original equilibrium state. If we let go of the pendulum, it swings back and forth from its disturbed state until it comes to a standstill again in its original equilibrium state. The pendulum in equilibrium is a typical example of an object that finds itself in a stable state."

"Do you have an example of an unstable state in material reality?" asked Ernest.

"The same pendulum also provides us with a good example of an unstable state," said the teacher. "Let's consider what mechanics calls a fixed pendulum. That's a pendulum on a pole instead of a rope. If we position the pendulum hanging vertically, there is one position in which the pendulum is at equilibrium, and that is when it is hanging exactly vertically. That is a typical example of an unstable state. In fact, if we disturb the pendulum, bringing it slightly out of this state, the forces that arise are such that the pendulum will go even further out of equilibrium and will never be able to return. The pendulum will start to swing and will gradually change to the stable equilibrium state that we described before. Hence this example of the perpendicular pendulum is an example of an unstable state."

"Right, I can see what you mean now," said Sonia, "and also how the state of the shopkeepers with their agreement to open for business for just three days a week is unstable. All that is needed is for their state to be disturbed slightly, for example by one of the shopkeepers not sticking to the agreement, and the forces that arise will set the whole network of shopkeepers in motion so that their joint state will shift even further away from their original state."

"The forces will very quickly involve all the shopkeepers in the change that has been brought about," said Ernest, "so that the network of shopkeepers will change to a new state in which they are once more open every day of the week."

"Quite so," agreed the teacher.

"Give us an example of a stable form of co-operation in social space," asked Jan.

"Well," said the teacher, "you have already provided an example yourselves. If the shopkeepers keep their shop open every day of the week, that is an example of a stable state."

"Why is that state stable?" asked Sonia. "Fundamentally it's not so different. One of the shopkeepers might decide to keep his shop open even longer, which will set off another change in state and a shift towards a new state."

"That's quite true, of course," replied the teacher, "but suppose for a moment that the Government has laid down the opening hours of shops by law. Further suppose that as a result it is illegal, for example, to open a shop on a Sunday. In that case, if one of the shopkeepers, all of whom now open their shop every other day of the week, nonetheless decides to open his shop on a Sunday, another set of forces would arise as a result of that action. Certainly, this shopkeeper will again make a profit in the short term, in accordance with the same mechanism as in the previous situation. However, since he is also committing an offence, the police will step in and, for example, fine him. This will mean that his profit will be converted to a loss. Even if the fine does not wipe out his profit completely, his brush with the law will harm him so much in the medium term that it would be a bad move all-round for the shopkeeper to open his shop on Sundays."

"So that means that the law can stabilise states that would normally be unstable," remarked Jan.

"That's right," said the teacher. "The introduction of laws has a stabilising effect. Many states of social entities that are unstable outside a State under rule of law become stable as a result of the existence of this rule of law."

"That's an important point," said Jonito. "I had not realised what a fundamental function the law could fulfil. Stabilising social entities that would be unstable without the law."

"There is another observation we can make," said the teacher. "In many cases the forms of co-operation we label as 'good' are actually 'unstable'. Anything good seems to move forward along the path of instability." The students sat up straight at this last assertion by the teacher.

"Good is unstable?" asked Sonia in amazement. "So does that make evil stable?"

"Well, there is certainly a connection between good and unstable, and between evil and stable," replied the teacher. "Let me try and illustrate that using an example. Let's compare the state of 'peace' with the state of 'war'. Peace, or in the case of co-operation we might call it a 'peaceful co-operation', is an unstable state. In fact, it is sufficient for one of the parties in a peace situation to violate that peace, and the forces that arise will ensure that the peace is further violated. The party that does not want to violate the peace situation is often compelled to do so out of sheer necessity in the interests of self-defence. On the other hand, a situation of 'war' is a stable situation. In fact, if one of the warring parties decides to disturb the state of war by opting for peace, no forces will arise naturally that allow the whole situation to evolve into a state of peace. On the contrary, this peace-loving person will come up against the natural forces that push him back into a state of war."

"That's a very pessimistic analysis," said Sonia. "Peace is unstable, while war is stable. If that's a fundamental truth, does it imply that evil must prevail because it is stable?"

"That's how it may look at first sight," said the teacher, "but the situation is not as simple as that. Let's just think about how mankind manages to 'preserve' and 'stabilise' peace. The first thing we need to mention is that at the end of the day no-one derives any good from a situation of "war of everyone against everyone". In the end, war is perceived as bad by all parties involved and that's exactly why it is associated with evil. One of the main reasons for this is that all-out war is a threat to older, previously acquired stabilities."

"Aha, I see what you mean," said Jonito. "Evil, though stable, is a threat to old, already stabilised instabilities."

"Right," said the teacher. "Just think of all the wrongs that happen during and as a result of war, and that are typical of war. Torture, betrayal, lies, rape, deterioration of faith, and so on, are all deteriorations of old, stabilised situations that could become unstable again as a result of the war. That is the first paradox we need to mention: although war and evil are generally stable, it is a stability that is a threat to other, previously acquired 'good' stabilities."

"How odd," said Jan. "That means that evil is only stable in the short term, since owing to the fact it is a threat to old stabilities, it too is doomed to disappear in the longer term. If a "war by everyone against everyone" is really being fiercely waged, eventually there will be no more people to keep the war going."

"Yes, you could put it like that," said the teacher, "but there is something to do with the difference between 'short' and 'long' term that we have to understand. We can emphasise the analysis we have carried out just a bit more forcefully, and that will clarify things slightly. Suppose one of you were to go outside now and rob the first passer-by you met on the street. Within a very short time, namely the brief moment in which the robber takes the possessions of the victim, there is a large personal material benefit for the robber. He has in fact got richer by acquiring the possessions of the victim. In the longer term, however, this benefit turns into a personal disadvantage. This can come about in various ways. There may be bystanders who rush to the victim's aid and overpower the robber, so that the robber's very short-term benefit is immediately turned into a disadvantage: he certainly loses what he has gained, and a lot more besides in that he is also punished. If there are no bystanders around to help the victim, it may be that the police are called in quickly and the robber arrested, so that his short-term benefit is still turned into a disadvantage. The chances of this happening are not particularly high, however—though not alarmingly small either—so that robbery can still be profitable for the robber. Another aspect that goes hand in hand with the act of robbery, and one that I would like to set against the 'short-term' aspect, is the way in which the robbery destroys certain stabilised social entities. In fact, by attacking and robbing the victim, the robber violates an unwritten agreement of co-operation, which says that people who walk the streets do not rob one another."

"But you can't talk about an avalanche effect here," said Sonia. "It's not as if this act of robbery is going to spread to other people on the street."

"Why is there no avalanche effect?" asked the teacher, and watched carefully to see who would answer him.

"That's quite obvious," said Jan. "If everyone were to start robbing people on the street, city streets would become intolerable."

"That's quite true," said the teacher, "but that's not sufficient reason. Real war makes normal life impossible for everyone, yet war often comes about as a result of the avalanche effect we talked about earlier."

"Yes, that's true," said Sonia, "so how come 'robbery' does not create an avalanche effect?"

"Well," said the teacher, "because the "walking-along-the-street-without-robbing-one-another" social entity is still, for the time being, a stabilised social entity. And that is partly due to the fact that 'robbery' is a relationship between two people, the victim and the robber, from which bystanders can hold themselves aloof. That is not the case with 'war'."

"That's why the short-term benefit is too risky as well," concurred Jonito, "because the whole of society is structured in such a way that an act of robbery gives rise to forces that work against the robber. In principle, the mechanism is probably the same as the one we described in the context of the shopkeepers and their keeping their shops open."

"Quite so," said the teacher. "In short, it is important for us to understand how the dynamic situations in social space we talked about can be stabilised."

"Stable—that sounds so sterile and dead," said Sonia. "I really have a problem with the analogies you are making. I get the impression that striving for stability cannot be the main objective. It's like the road to everything that's sterile and dead."

"Very good point, Sonia," said the teacher. "There is indeed a link between stability and death. The example of 'peace' and 'war', and their clear similarities with 'instability' and 'stability', is slightly misleading. To clarify things a bit more, let me use another analogy from the material layer of reality. You all know that matter is organised in different ways. Broadly speaking there are four easily distinguishable ways in which matter is organised: gases, liquids, crystals and living matter."

"That's interesting," said Ernest. "How are these different ways distinguished from one another?"

"That's just what I want to talk about now," said the teacher. "and we shall see that when we understand it clearly, it will shed new light on each of the questions we asked ourselves about social entities and about the ways in which these can stabilise?" The waiter appeared and they ordered another round of drinks. The students were paying close attention because they felt that the conversation was getting closer to the essence of the problem. The teacher waited until the waiter had come back with the drinks before resuming his story.

DIFFERENT TYPES OF ORGANISATION: LIFE AND DEATH

"As we have already mentioned, physical space is homogeneous and isotropic," continued the teacher, "and offers a variety of possible stable forms of movement. The idea of equality is nicely realised through this homogeneity and isotropy, as it makes the places of all moving objects potentially equivalent. It is in no way necessary that objects try to occupy each other's place, which in any case is impossible because of physical laws. Nevertheless they all manage to bring about stable patterns of co-operation, in which it is essential that they are fundamentally 'distinct' from one another. Yet places are not monopolised either. The stable pattern in which the objects find themselves is a dynamic pattern where movement is essential, and where this movement implies that objects can occupy one another's place, one object taking up the place that another has vacated, without, however, coming into conflict with one another. Only these conclusions might give us inspiration for a dynamic development of social space, which permits and fosters stable movement patterns. It becomes even more interesting, however, if we note that this matter has stabilised itself in fundamentally different ways within these possible dynamic movement patterns. Let's take a closer look at these different ways. First of all there are the gases. These are entities made up of atoms and molecules that interact mainly by just hitting each other continuously. Then there are liquids. These are also entities made up of atoms and molecules, but which interact as one imagines what happens with a box full of frogs rolling and slithering over one another. The interaction between the various atoms and molecules in a liquid is not really giving rise to constructions. In the case of gases and liquids, one could not talk about 'constructions' of atoms and molecules. Both types of organisation are collections of atoms and molecules and are not constructions. Then there are crystals. Here we should perhaps first observe that all 'solid substances' are made up of crystals, with the exception of a few substances that are liquids, but which set so hard that they appear to us to be solids."

"Normally people think of splendid precious stones when they talk about crystals," said Ernest.

"That's true, but it's misleading," said the teacher. "Those superb stones, such as quartz, ruby, sapphire, diamond and many others, are highly pure crystals and are simply made up of a few kinds of atoms and molecules. Diamond, for instance, is a crystal that is made up purely of carbon atoms. All solids are composed of crystals; even metals are crystals. Many solids, such as most stones, are not precious stones or semi-precious stones since they are a disorganised mixture of many different crystals, while others, such as the metals for example, are composed of just a few kinds of atoms or molecules, but because of the way in which these specific atoms or molecules interact they look completely different from precious stones."

"Which liquids do we think of as being solids?" asks Sonia.

"One example is glass," replies the teacher. "Glass is a liquid that sets so hard that it appears to us to be a solid. That is also the reason why a pane of glass placed in a frame will gradually flow downwards. Window panes that have been in a frame for a long time are thicker at the bottom than at the top. Pitch is another example of a liquid that looks like a solid to us. But we are getting away from what I wanted to tell you. The point we should make is that crystals are actually constructions of atoms and molecules. Crystals are made up of atoms and molecules according to particular symmetrical and geometric patterns. There are a few basic patterns that are repeated again and again, eventually culminating in a gigantic construction."

"You persistently use the metaphor of a 'construction' when talking about crystals," said Jonito.

"I have done that deliberately," said the teacher. "Like constructions, crystals are made of basic entities, atoms and molecules, put together according to a few simple building principles, based on symmetries and repetitions."

"How is the fourth type of organisation you mentioned different from the others?" asked Ernest.

"Well," said the teacher, "I would like to come on to that now. The fourth type of organisation, living matter, is not a construction. It is a form of co-operation."

"That's an interesting idea," said Jonito. "Living matter as a form of co-operation."

"Once again I would like to draw your attention to the fact that these comparisons are metaphors, in principle," continued the teacher. "Nevertheless we can still find out a few fundamental things in this way. A living cell organised around the macromolecule that we now call DNA is an association of macromolecules that we call proteins. It is the co-operation between all these macromolecules that stabilises the cell and gives it life. This form of co-operation experiences the same stability problems as the social forms of co-operation we mentioned earlier. If some of the proteins no longer fulfil their role in the co-operation, the cell becomes sick and may possibly die. Multicellular living beings are, in turn, forms of co-operation between all cells and the macromolecules. This co-operation is participatory and is not controlled from a central point. The co-operation between the sub-entities of living matter takes place in a self-organising way. There are no specific rules that are fixed once and for all, but instead general principles that can stabilise the process of self-organisation. Living matter is a form of co-operation that brings about a stabilised unstable situation."

"Fascinating," said Ernest. "I am starting to see where you are heading. So is it the case that self-organising social entities can be regarded as the continuation of life?"

"And hierarchical social entities as a division of life, so that they are actually dead, like crystals," added Jonito.

"You are on the right track," said the teacher. "Life has always tended to move in the direction of the instability of forms of co-operation and then stability, while at the same time creating space for this stabilisation. Look at the diversity of life. Life has not let itself be contained in narrow, normative forms of space, but instead has created a highly diverse presence."

"It is precisely this diversity that is required for the survival of life on Earth," said Jonito. "That fact has already been recognised."

"And this diversity also serves as the starting point for the Darwinists' theories about natural selection as a mechanism for evolution," said Jan.

"Quite so," said the teacher. "The Darwinists base their theories on a situation that is already highly diverse. They forget, however, that this diversity must arise in order to stabilise the original instability of the early forms of co-operation between macromolecules. Diversity is not just a magnification of space as a result of chance mutations, which then permit natural selection; it is also necessary for the stabilisation process of each individual form of co-operation in itself. Environmentalists have already realised this, though in a more intuitive way, without clearly understanding the basic principles."

"Why is diversity an essential element in this type of organisation?" asked Jan.

"In order to understand that, we have to look at this type of organisation more closely," said the teacher. "Incidentally, we can take the emergence of forms of co-operation going on in social space as an example. We can see that different types of co-operation are being tried out there. Some of these are strictly hierarchical, a good example being the structure of an army or a police force. These hierarchical forms of co-operation are organised in a way that is more like that of the crystal in physical space. There are simple principles, generally based on insights of symmetry and geometry, which externally describe the way an organisation functions. This must be the case, because in a hierarchical type of organisation the underlying sections must be known, overseen, controlled and directed by the hierarchical superior of this underlying section. An intrinsic feature of this hierarchical type of organisation is that it can be described as cognitively reductionist; otherwise the hierarchical superior would be unable to fulfil his task properly. Hierarchical types of organisation therefore have no need for the emergence of new regulating principles, over and above the rules and regulations that have been introduced, but which, in principle, can be known and controlled by the most senior person in the hierarchy, and used by him to direct the whole organisation. Clearly we are talking about a pure hierarchy here, something that is never found in society. The structures of an army or a police force are actually attempts to come as close as possible to this hierarchical type of organisation."

"But as far as I am concerned, diversity is not an essential ingredient here," remarked Ernest.

"That's quite true," concurred the teacher. "A hierarchical type of organisation does result in a certain diversity, but it is a diversity that has to do solely with the various 'structures' that can be erected, based on the primitive behavioural patterns that go with hierarchical behaviour. We could try to give as full a description as possible of the elementary behavioural patterns that go with the hierarchical type of organisation: these include elements such as 'obeying one's hierarchical superior', 'giving clear orders to one's hierarchical inferior', 'planning the collaboration between one's hierarchical inferiors to the letter and in as much detail as possible'. Incidentally, that's why the diversity of the world of dead solids, i.e., crystals, is limited, certainly when we compare it with the diversity of living matter."

"I still don't understand what diversity—as a necessary state—is the result of," said Jan.

"We can understand the creation of diversity as follows," said the teacher. "Natural selection, the Darwinian mechanism that brings about evolution, starts out from a situation in which there are various entities from which a selection can be made, and a mechanism of mutation that allows various new entities to be created that will contribute to the process of natural selection. This Darwinian analysis is generally based on a given moment in time, or more accurately a phase in the whole development process, a phase that is historically determined by the main area of interest of Darwin himself, namely the evolution of the 'species'. This is a phase in which various entities already exist, entities that are diverse and compete with one another in the fight for the 'survival of the fittest', and in which a mechanism—for the specific Darwinian situation, the mechanism of mutation—exists, which can arbitrarily give rise to new entities. The phase that we particularly wish to examine here in the context of social entities is a slightly, though not completely different phase, as we shall see immediately. What we are talking about here are entities that are forms of co-operation between individual human beings. If we wanted to find an 'older' analogous phase, such a phase would have more in common with the phase in which different living cells started working together to bring about stable multicellular organisms. By definition there will be a lot of opportunities for establishing stable forms of co-operation between individual human beings, and it is this brutal fact of the 'under-determinateness' of the demand for stability that will bring about diversity as a natural consequence. This diversity is still growing, which explains the possibility of diversity. The necessity of diversity lies in the buffer effect of the network of diversified entities. Think, for example, of the food chain in a biotope, where populations start to fluctuate if the diversity of species in the biotope decreases. It is simply the case that every new form of co-operation is more robust, as a new type, if it hedges its bets. The situation of different variations of this type will have a better chance of becoming stable since these variations are better able to cope with all kinds of external destructive forces."

"So there is competition and natural selection," said Jonito.

"Quite so," said the teacher. "The differentiation between these two phases is merely to clarify matters. These two mechanisms, under-determinateness, which gives rise to natural diversity, the kind of diversity that gives rise to robustness, and the 'survival of the fittest', which gives rise to selection, are both still at work. There is simply a shift in emphasis in the two different phases. I also wanted to mention them because it enables us to some extent to broadly outline which mechanisms will play a key part in our society in the near future."

"That's interesting," said Jan. "Which mechanisms do you specifically have in mind?"

"The situation in which we now find ourselves," continued the teacher, "i.e., under-determinateness of the stability conditions, which mainly keeps the mechanism of the creation of natural diversity going, will result in more and more different and diverse types of co-operation being developed. The first signs of this are already evident. Often, if there are relevant examples, these major forms of co-operation are called sub-cultures. More and more diverse sub-cultures will surface, leading to a very diverse range of possible forms of society. This is the real expansion of social space, which will become increasingly large. Individuals and groups will be able to choose between various forms of society, will be able to occupy different places in this social space, and, most important of all, bearing in mind our previous analysis, individuals and groups will move dynamically through this social space. Everywhere they will be able to integrate themselves within a different form of society, and all these diverse forms of society will be stable compared with the disintegrating social forces we mentioned earlier. Not one of these forms of society, and this will by definition be true, since otherwise they simply will not arise in social space, will be subject to these disintegrating forces. The opportunities presented by freedom, which, if just one or a few forms of society exist, have either had to be curbed in the case of most individuals, or were subversive and consequently destructive, will now be turned into opportunities for movement, with people 'surfing' between the various types of society offered."

"So, freedom becomes freedom of movement?" asked Jonito.

"Exactly," replied the teacher. "This will be one of the most unexpected consequences, and will mean huge changes for today's society. There is evidence that these changes are already well under way."

"Can you just explain that a bit more?" asked Ernest.

"Well," continued the teacher, "as social space expands, the diversity of the places offered in this social space will increase. These places that are offered are none other than the various forms of society we talked about. Individuals will be able to move in this social space, which means that they can look for various forms of society. Each of these places in social space will be stable, which means that each of the forms of society offered will be internally armed against the disintegrating forces."

"Do you mean that each of the forms of society offered will be governed by laws and regulations, whereby situations such as the 'prisoners' dilemma' and 'the war by everyone against everyone', as a result of the destructive forces of envy, will be impossible?" asked Jan.

"Exactly," said the teacher. "This will, by definition, be the case since social space will comprise places that are offered to 'stable forms of society'. Just as physical space comprises places that are offered to stable physical systems."

"I get it," said Ernest. "Forms of society that are not stable will, by definition, not be present in this social space."

"Yes, that's it," concurred the teacher.

"And the freedom of the individual will be that he can choose again and again to shift from one form of society to another," added Jonito.

"Quite so," said the teacher. "The natural course of life of an individual or a group of individuals will be to travel, as it were, between the various forms of society offered and move thus within social space. In each place one will find oneself in a form of society governed by other, different stabilising laws and regulations. One lives within this form of society if one obeys the laws and regulations. There are no restraints placed on individual freedom, however, since one can decide at any time to move to another place in social space and choose a different way of living. And diversity will lie in the enormous number of different forms of society that will co-exist."

"This scenario implies that all forms of society are open to everyone," remarked Sonia.

"True," said the teacher, "and here we can see that the analogy with physical space sheds yet more light on the subject. I mentioned the isotropy and homogeneity of physical space, and it is precisely these properties that are linked with openness. Social space will be fundamentally isotropic and homogeneous, which means that every place has the same potential with respect to each individual or group of individuals. This isotropy and homogeneity therefore implies a particular form of 'openness' for the 'empty' social sphere. It is always possible to try a new form of society. There is certainly room for it."

"On the other hand," remarked Jan, "I remember that you also came to another conclusion. One of the useful analogies with physical space was the one that showed us the destructive power of envy. Why the 'place' of an entity could not be occupied by another entity in physical space, even in a dynamic situation. I think there's a contradiction with openness as a fundamental option. This tends more towards closedness."

"A very accurate observation," said the teacher. "There is indeed a fundamental closedness that has to do with the individual character of the individual or the group of individuals. The fundamental openness we talked about is that of the 'empty' space, the potential place for no matter what type of society. Many forms of co-operation are open in terms of their environment, but closedness is also an option. If we do something 'stable' in physical space as a group, for instance go for a walk with a few friends, there will be openness with respect to the way in which people take part in the walk. At the same time, however, though not in contradiction—and that is an important point—will each individual retain most of his or her individuality during the act of co-operation that the walk constitutes? The same combination of environment-openness and identity-closedness will probably be useful states in social space. We can already see examples of this, incidentally."

"What examples are you thinking of?" asked Ernest.

"Even the competitive capitalist market has basic built-in mechanisms to protect information relating to the identity of, for example, a company," said the teacher. "But on the other hand, there is also a fundamental environment-openness."

"That's true," remarked Jan. "It's prohibited to use the name or logo of an existing company, but it is permitted to operate within the business environment of that company."

"That is one example," said the teacher.

"Nevertheless there will be competition between the various forms of society," said Sonia, "so that they will still threaten one another's place."

"Quite so," said the teacher, "and that is a fundamental observation: there will be competition between the different forms of society. Incidentally, this kind of competition has played an important part in our history, and has often deteriorated into war, with individuals who were part of one form of society being killed or taken prisoner. They were also often forced to become integrated into the victorious form of society. In this context there is an important point I would like to make. This competition between forms of society, which leads to war, death and enforced integration, will take place on a grand scale. There is obviously a consensus on the fact that this is a 'bad' thing. It might come about, therefore, that if social space is big enough, this specific form of competition might no longer have a place."

"That seems very utopian to me," remarked Jan. "Let us just go back to another analogy that you have often made, namely that of living matter, animals and plants as forms of co-operation between cells and macromolecules. It is evident that competition, with all its 'bad' consequences, is still raging fiercely in this area. There are animals—I am referring to animals of prey—that can only survive by destroying other animals. Other animals survive by destroying plants. Things do not look so good, even though this layer of reality, the biological layer, has had millions of years in which to form. Is it not highly unrealistic to hope that the competition between social ways of living will proceed any differently?"

"You are absolutely right," said the teacher. "Nature is ruthless, as it is so often said. Living matter survives by 'eating' other forms of living matter. In the same way the various forms of society will certainly 'survive' by 'eating' other forms of society."

"That seems to me to be a terrible vision of the future," said Jan. "OK, humans will have a wide range of diverse forms of society from which to choose, and between which they can move, but at the same time they will be torn apart by ruthless competition and war between these different forms of society." A long silence fell and the teacher looked at the circle of intent students.

"You are right," he said. "Evil will not be driven out by the evolution I am outlining, however irrevocable this may be. But there is one aspect we are overlooking. If one animal eats another, this is bad news for the animal that is eaten. If ice melts and becomes water, this is bad news for the ice. If one form of society absorbs another, this will be bad news for the society that disappears. It is noticeable that we do not feel that it is 'bad' in the example of the ice, but we do in the example of the animal. What might the reason for that be?"

"Because we are much closer to the animal than to ice," said Sonia.

"That is one aspect," said the teacher. "But there is another one that I would like to discuss. If one animal is eaten by another, is this bad for the macromolecules of the eaten animal?" Again there was silence as the surprised students thought about the teacher's question.

"Aha," said Jonito, "I understand what you mean. When animals eat plants or other animals, the basic constituents—i.e., the atoms and molecules—are not destroyed. When these are eaten they are simply moved to another entity of living matter. So they are not really being destroyed."

"Exactly," said the teacher. "Animals or plants that live off other animals or plants do not, as a rule, destroy the basic constituents of these animals or plants. It is much more efficient, and hence more competitive too, to use these basic constituents in their own structure of living matter."

"It is Darwinian competition itself that has most likely ensured that this is a better strategy than destruction," argued Ernest.

"You are probably right," said the teacher. "So it is reasonable to expect that the basic constituents of forms of society, i.e., individuals or groups of individuals, will be respected just as much in this competition between different forms of society."

"I see where you are heading," said Jan. "If that were the case, competition between different forms of society would occasionally result in a large group of people abandoning a particular form of society in favour of another."

"This kind of migration would happen," said the teacher. "Migration can always be looked at from two points of view. From the viewpoint of the individual or group of individuals who emigrate to a 'better' form of society, or from the viewpoint of the forms of society themselves, where one of the two forms of society is destroyed by the other."

"In the same way as you might consider the fact of a plant being eaten by an animal from the viewpoint of the macromolecules," said Jonito. "The simple macromolecules of the plant would then get the opportunity, as a result of this migration, to play a part in the form of co-operation between the much more complex animal macromolecules."

"Quite so," said the teacher. "It depends on how you look at it."

"I am starting to get a picture of your vision of the future," said Jan. "I have another question, however. You argue that only those forms of co-operation that are stable will exist within the new social space. Does this mean that there will be no more free-riders, no prisoners' dilemma and no more envy? Will the individuals who cause this kind of destructive force simply be 'eliminated'? Does not sound like much of an outlook for those individuals."

"Well," said the teacher, "one could argue that individuals are not present in social space anyhow. Social space is, by definition, the sphere where forms of co-operation between individuals have found a place. Once again 'by definition' this social sphere will be occupied by forms of co-operation that are stable, and in which people have solved this problem of 'destruction'. It is not so clear what will happen to individuals who set these destructive forces in motion. Again we can look at the material layer to provide us with some answers. The molecules of which matter, including living matter, is composed are not present in physical space either. This property is called 'non-locality' [36]. If a photon travels from one macroscopic entity to another, en route it is not present in space, and has the freedom that it once possessed before physical space shaped itself as a theatre for the forms of co-operation of macroscopic material entities. In the same way I suspect that once a well-defined social space has been created, individuals will not lose their freedom. While they travel from one form of society to another they will simply not be present in social space."

"Aha," said Jonito, "does this mean that you think that social space is more likely to have a quantum structure? What I mean to say is, If we had to build a mathematical model of this social space, would it be a quantum model?"

"That will depend on the level of the description," said the teacher. "Let's consider for a moment what happens in physical space. The classical Newtonian description of physical space is not quantum. Only the movements of the forms of co-operation—in this case the macroscopic physical entities—are described. If one also wants to describe the behaviour of the individual entities, i.e., the atoms and molecules, one comes up against quantum mechanics."

"I would like to come back to the link with the paradigm shift from observational to participatory society," said Sonia. "Do we understand this any better now?"

"Yes," said the teacher, "it's time to look at that aspect again."

THE MISSING INTERACTIONS: OBSERVATIONAL AND PARTICIPATIVE WORLD VIEW

"In an earlier discussion we talked about the transition from an 'observational society' to a 'participatory society'," said the teacher [1]. "I would now like to try and fit this transition into the evolution from a static form of society to a dynamic form of society, in order to gain a better understanding of the participatory deficit, the greatest excrescence being the scourge of unemployment in the late twentieth century." The students listened, fascinated, now that the teacher had broached the subject with which the discussion had begun. The teacher continued:

"We have already mentioned that the transition from a static to a dynamic society led to great upheaval due to the fact that old mechanisms to combat the destructive forces were no longer efficient. We have also clearly seen how social space was created at the same time as this transition from static forms of society to dynamic forms of society. We have also analysed how the 'law' acts as a stabiliser of the new forms of society that are reluctantly learning to move in this social space [41].

We have often used analogies from physical space to explain this construction process within social space. For instance, we have analysed the way in which macroscopic physical entities stabilised on the basis of the current physical laws. We have seen how living matter, which we can regard as a 'form of co-operation' between macromolecules, is distinguished from other forms of organisation of matter, i.e., crystals, liquids and gases, which we can classify as being 'dead'. We can now say that the physical entities 'interact' within physical space and it is through these interactions that they become stabilised. A very special act of creation occurred, however, during a particular phase in the evolution of living matter. Animals and their descendants, people, developed 'eyes'. We should consider this fact for a moment if we want to grasp the problem of participation more clearly. What is the main function of 'the eye'?" The teacher paused for a moment and watched the inquiring expressions on their faces.

"To make a model of physical reality," said Jonito.

"That's quite right," said the teacher. In fact, the eye does something very special if you classify it among interactions that frequently take place between physical entities. The eye interacts with the surrounding physical entities in a specific way that is different from the types of interaction that underpin the stabilisation of the physical entities. The eye primarily wants to 'observe' and therefore tries to interact in such a way that the 'effect' on the 'observed' physical entity is minimal. It wants to cause as little 'disturbance' as possible in order to be able to understand the individual reality of the physical entity studied. This is an interaction—let us call it an 'observation'—that is not aimed at stabilising the physical entities. Within the context of an overall evolution, observation appears much later on the scene, namely during the phase when the variety within physical space has already increased enormously and Darwinian competition and selection are playing high. In this phase it is definitely an advantage to have an 'observed picture' of the surrounding reality, on the basis of which 'predictions' can be made."

"Whereas the interactions that are necessary for the stability of the physical entities are of a different type?" asked Jan.

"Yes, this is clearly the case," said Jonito. "The typical interactions that bring about stability are usually symmetrical. It is not the case that one of the interacting entities observes the other, forms a picture of it and then 'decides' on the basis of this picture which reaction will be sent to the observed entity."

"Quite so," said the teacher. "Typical interactions, such as those between macroscopic physical entities in classical mechanics and those between microscopic entities in quantum mechanics, are not of the 'observational type', as is the case with the 'eye'."

"I think I see where you are heading," said Sonia. "On account of the fact that this 'observation' has been crucial for us in acquiring most of the knowledge we have accumulated about reality, the 'observation' type of interaction is a very powerful archetype for us. Is that perhaps the reason why we are constantly tempted into taking an 'observational world view'?"

"Yes, I think that is in fact the main reason," said the teacher. "I have to say, however, that by putting forward the special 'observation' interaction as being important to us because of the development of the eye, I am probably making this observation interaction too specific. It is indeed a very important type of interaction in itself, in that, on the one hand, it disturbs the observed entity as little as possible and, on the other hand, it enables as much information as possible to be gathered about the state of the observed entity. This observation interaction can be distinguished from other types of interactions in a very formal way [42]. It is quite possible that quantum entities that interact with one another have already used it, though current quantum theories provide no definite answers on this point. In any case the interactions that are important for the stability of the co-operation between entities are of a different type."

"So is that why it's so difficult for us to intuitively grasp the nature of these stabilising interactions?" asked Jan.

"Exactly," said the teacher. "We have now reached the stage where we can express our criticism of the observational world view, which we formulated intuitively in our earlier discussion [1], in a more structured way. During the process of construction of social space, the construction of the interactions required to stabilise these social entities was also taking place. The process of forming these interactions in our present-day society is fully under way: it is the legislative, executive and judicial powers that play a principal part in this. These powers largely operate, however, on the basis of the observational world view. They implement regulations and laws by reasoning on the basis of 'observational' models of social entities. The laws are largely based on an observational view of the social entities in society. It is very difficult to capture a truly dynamic situation with this."

"What do you mean?" asked Ernest.

The teacher continued: "The stabilising interactions are not observational from one entity to another because they require 'continuous' feedback about the movement of one entity towards the other. If one tries to capture such a situation of continuous feedback in a model that is observational—in other words where one entity is observing the other—

one comes up against serious problems. It can be compared with the attempt to integrate by summation."

"I get it," said Jonito. "Only continuous mathematical models can describe this continuous feedback. Is that also perhaps one of the reasons why the stabilising interactions of material entities in physical space can be readily captured in mathematical models?"

"Yes, that's right," said the teacher.

"And that's why the complex structure that stabilises a truly dynamic situation cannot be described using an observational approach," added Sonia.

"You could put it like that," said the teacher.

"Is that also the reason why you claimed in our last discussion that ethical forces can bring about this stabilisation of social entities, whereas cognitively formulated rules and laws, i.e., those resulting from the observational world view, cannot?" asked Sonia.

"Quite so," replied the teacher. "We have to be aware that the problem of the construction of the stabilising interactions was not just posed within a dynamic form of society. It is simply more complex within this kind of dynamic form of society, but it was already present within the static form of society in primitive societies, where these stabilising interactions took the form of rituals and customs and were expressed in ethical structures. Ethical structures ensure that destructive, disintegrating forces do not hold sway. These ethical structures were often, and still are, to a great extent implicit. But in this case implicit also means 'not purely observationally interactive'."

"What does this mean in concrete terms for our present-day society?" asked Sonia.

"Let's just go back to the issue of unemployment, which was one of the fundamental expressions of the participative deficit in twentieth-century society," continued the teacher. "In the first place we can say that there was certainly no 'shortage' of work. There was a tremendous amount of work available. The real problem was that all this work was in an area of social space where the interactions that were maintained by the free market economy could not penetrate. The work available was mainly in that area of social space where mobile individuals or groups of individuals had appeared, and was often associated with quality improvement, for instance in the welfare sector, health care, ecology, and so on. Often, however, it involved efforts to solve harrowing human problems in less developed areas: starvation and malnutrition, water supplies, etc. The interactions that were maintained by the free market economy, and that were chiefly based on the exchange of material goods, could not penetrate into these areas of social space. Incidentally, this fact was recognised by many philosophers at the end of the twentieth century, but they failed to draft regulations or laws that would bring about the interactions that would have made co-operation in these areas of social space possible. The main reason for this was that regulations and laws were still constructed on the basis of the observational world view. There is an urgent need to introduce interaction patterns into our present-day society that make it possible to involve this area of social space—where there is now sufficient work, but where the present interactions of the free market cannot penetrate—in the total stabilising dynamics of society. How we might go about this could form the subject of another discussion, since a number of steps have already been taken in this direction." It was now very late and the teacher and his students decided to call it a day and keep this subject for their next meeting.

Diederik Aerts
Center Leo Apostel
Brussels Free University
Brussels, Belgium
e-mail: diraerts@vub.ac.be

REFERENCES

[1] **Aerts, D.**, "Participating in the World: research and education in a changing society", in: **Aerts, D., Gutwirth, S., Smets, S. and Van Langenhove, L.** (eds.), *"Science, Technology and Social Change", the Orange Volume of Einstein meets Magritte*, Kluwer Academic Publishers, Dordrecht, 1999.

[2] The concept of 'physical space' was analysed in this way by Henri Poincaré in [3,4], and later used productively in the theory of relativity. See also the detailed analysis in [5, 6, 7].

[3] **Poincaré, H.**, *La science et l'hypothese*, Flammarion, Paris, 1968.

[4] **Poincaré, H.**, *La Valeur de la Science*, Flammarion, Paris, 1948.

[5] **Aerts, D.**, "The construction of reality and its influence on the understanding of quantum structures", *Int. J. Theor. Phys.*, **31**, 1992, p. 1815.

[6] **Aerts, D.**, "Framework for possible unification of quantum and relativity theories", *Int. J. Theor. Phys.*, **35**, 1996, p. 2431.

[7] **Aerts, D.**, "Relativity Theory: What is Reality?", *Found. Phys.*, **26**, 1996, p. 1627.

[8] René Girard examines the role of the force of mimetic desire in detail in [9,10,11,12].

[9] **Girard, R.**, *La violence et le sacré*, Grasset, Paris, 1972.

[10] **Girard, R.**, *Des choses cachées depuis la fondation du monde*, Grasset & Fasquelle, Paris, 1978.

[11] **Girard, R.**, *Le bouc emissaire*, Grasset, Paris, 1982.

[12] **Girard, R.**, *La route antique des hommes pervers*, Grasset, Paris, 1985.

[13] Most studies of the social insects take instinctive behaviour alone as the starting point for the whole of social behaviour [14,15,16,17,18], which does not mean that no elements on a non-instinctive nature might be present here too.

[14] **Wilson, E.O.**, *Sociobiology: the new synthesis*, Harvard University Press, London, Cambridge, 1977.

[15] **von Frisch, K.**, *The dance language and orientation of bees*, Belknap Press of Harvard University Press, Cambridge, 1967.

[16] **Moritz, F.A.R. and Southwick, E.E.**, *Bees as superorganisms*, Springer-Verlag, Berlin, 1992.

[17] **Lindauer, M.**, *Communication among social bees*, Harvard University Press, Cambridge, London, 1971.

[18] **Bourke, A.F.G. and Franks, N.R.**, *Social evolution in ants*, Princeton University Press, Princeton, 1995.

[19] For a study of the social behaviour of other social mammals, we refer to [20,21,22,23].

[20] **Chauvin, R. and Ordish, G.**, *Animal societies from the bee to the gorilla*, Gollancz, London, 1968.

[21] **Schaller, G.B.**, *The mountain gorilla: ecology and behavior*, University of Chicago press, Chicago, 1963.

[22] **Wrangham, R.W.**, *Chimpanzee cultures*, Harvard university press, Cambridge, 1994.

[23] **Klinghammer, E.**, *The behavior and ecology of wolves*, Animal Behavior Society and Symposium on the Behavior and Ecology of Wolves, Garland STPM Press, New York, 1979.

[24] **Aristotle**, *Ethica*, English translation, *The Ethics of Aristotle*, Penguin, Harmondsworth, 1953.

[25] **Aristotle**, *Politica*, English translation, *The Politics*, Penguin, Harmondsworth, 1962.

[26] **Plato**, *L'Etat ou la République de Platon*, traduction de Grou, Emmanuel Bekker, Lefevre et Charpentier, Paris, 1840.

[27] **Hobbes, T.**, *Leviathan*, Everyman's Library, London, 1983.

[28] **Poincaré, H.**, *La Science et l'Hypothese*, Flammarion 1963.

[29] **Achterhuis, H.**, *Het rijk van de schaarste*, Ambo, 1988.

[30] **Newton, I.**, *Les Principia de Newton*, par Michel Blay, Presse Universitaire de France, Paris, 1995.

[31] The nature of the indeterministic aspect of quantum mechanics is analysed in [32], [33], [34], [35].

[32] **Aerts, D.**, "A possible explanation for the probabilities of quantum mechanics", *J. math. Phys.*, **27**, 1986, p. 202.

[33] **Aerts, D.**, "The origin of the non-classical character of the quantum probability model", in: **Blanquiere, A., Diner, S. and Lochak, G.** (eds.), *Information, complexity, and control in quantum physics*, Springer-Verlag, Wien-New York, 1987, p. 77.

[34] **Aerts, D.**, "Quantum structures due to fluctuations of the measurement situations", *Int. J. Theor. Phys.*, **32**, 1993, p. 2207.

[35] **Aerts, D.**, "Quantum structures: an attempt to explain their appearance in nature", *Int. J. Theor. Phys.*, **34**, 1995, p. 1165.

[36] A detailed description of the interpretation of non- locality presented here can be found in [37], [38], [39], [40].

[37] **Aerts, D.**, "An attempt to imagine parts of the reality of the micro-world", in the proceedings of the conference *Problems in Quantum Physics*, Gdansk '89, World Scientific Publishing Company, Singapore, 1990, p. 3.

[38] **Aerts, D. and Reignier, J.**, "The spin of a quantum entity and problems of non-locality", in the proceedings of the *Symposium on the Foundations of modern Physics*, Joensuu 1990, Finland, World Scientific Publishing Company, Singapore, 1991, p. 9.

[39] **Aerts, D. and Reignier, J.**, "On the problem of non-locality in quantum mechanics", *Helv. Phys. Acta*, **64**, 1991, p. 527.

[40] **Aerts, D., Durt, T. and Van Bogaert, B.**, "Quantum Probability, the Classical Limit and Non-Locality", in: **Hyvonen, T.** (ed.), the proceedings of the *International Symposium on the Foundations of Modern Physics*, Helsinki 1990, Finland, World Scientific, Singapore, 1993, p. 35.

[41] **Aerts, D. and Gutwirth, S.**, *The dance of law*, in preparation.

[42] **Aerts, D.**, "The origin of the non-classical character of the quantum probability model", in: **Blanquiere, A., Diner, S. and Lochak, G.** (eds.), *Information, complexity, and control in quantum physics*, Springer-Verlag, Wien-New York, 1987, p. 77.

KYRA MARIE LANDZELIUS

HUNGER STRIKES: THE DRAMATURGY OF STARVATION POLITICS

Hunger strikes rearrange the geographies of embodiment: transfiguring the fragility and destruction of life 'within' the experiential, phenomenologically discrete subject into a performance of life philosophies projected 'beyond' a politico-moral agent. This conversion expresses a multiplicity of contrasting discourses: rhetorics of domination and empowerment, symbols of the body colonized and ransomed, the body reified and defiled; and amplifies them into a spectacle for public scrutiny, a spectacle insistent in its imagery and immediacy. What is the message of this dark passion play? How does corporeal rhetoric challenge the discursive practices of power? How does the wastage of starvation conjure a sweeping indictment of authority, a plea for revision of community ethics and the foundations of morality? Inspired by these questions, this paper will consider some of the dramaturgical qualities and paradoxical representations in the manifestation and moralism of hunger striking.

SCENES & SETTINGS

In the theatrical politics of the twentieth century, hunger strikers have staged extremist rebellions for a multiplicity of 'causes': ranging from wholescale indictments of totalizing powers to more localized claims for citizens' entitlements within late capitalist consumer democracies. As the last resort of the disenfranchised, protest fasting has wielded a subtle but persistent presence in the modernist battles against colonial and communist regimes, as well as post-colonial and post-communist dominations. It has fought for women's suffrage, self-determination, civil liberties, prisoners' rights, and like struggles in the campaign for representational, inclusive governance. Additionally, self-starvation has been a weapon in less notorious, less grand missions (if perhaps no less noble, or sweeping in their interrogation of the fundamental operations of power). In the past decade alone, this hodgepodge of motives has included: homelessness, environmental pollution, immigration laws, work conditions, foreign policy, military activity, government fraud, and a miscellany of personal tragedies—all perceived as abuses of power or abhorrent symptoms of its gross negligence. Like every human endeavor, these issues are inescapably bodily concerns; and yet it is with a certain shock and

© 1999 *Kluwer Academic Publishers. Printed in The Netherlands.*

surprise that we recognize the starving body of the political faster as the battleground for social reformation.

A few vignettes of the hunger strike event may evoke the imagery and intrigue of this curious phenomenon. The plights of Gandhi, Mandela, Bobby Sands, and Sakharov are familiar to many; and the showdown at Tianamen Square still burns in memories, while the sagas of the British suffragettes, the Hunger marchers of Britain's unemployed, the Mexican tin miners, Israeli doctors, Pakistani immigrants in France, Soviet enemies of the state, and countless others may not spark recollection [1]. There are numerous untold stories as well. A Canadian farmer's hunger strike directly inspired the Ottawa government to rewrite finance legislation to protect farms against farm foreclosure [2]. A bisexual judge in Chicago carried out a 58-day fast against bias toward homosexuals. Russian-Canadian separatist Mary Astoroff starved to death to resist cultural assimilation [3]. In 1986, astrophysicist Charles Hyder deployed a fatal hunger strike to protest the arms race; and recently Boston lawyer Jennifer Harbury fasted to gain C.I.A. assistance in tracking her missing Guatemalan husband [4]. Survival for peace activist, Jerry Rubin, upon ending his fast to protest media violence, sent a 'peace pizza' to actor Sylvester Stallone of 'Rocky' fame [5]. As I write, Chinese women held in a California detention center engage hunger strikes to proclaim childbearing as a basic human right, denied them under China's draconian family planning. The list goes on ... Then there's the twisted tale of Jan Smith, a Wall Street broker who lost his self-made fortune in the market crash of 1988, and undertook a hunger strike to call attention to his impoverished plight. But, oddly and fascinatingly, even Smith's tragicomedy garnered public support from a group of activists who picketed on his behalf, parading around the cardboard box in which he had staked himself in the financial district of New York. Indeed, not uncommonly have hunger strikes inspired a sort of 'groupie' contagion, a mimicry phenomenon wherein onlookers have undertaken solidarity fasts in show of support for the striker's cause.

In some cases the hunger strike has been a mere 3-day fast, in some cases—as with the fate of astrophysicist Charles Hyder—the individual starved him- or herself to death. So what is happening here? And what is this happening called a hunger strike?

A historical appraisal of hunger striking configures an intriguing pastiche: with a self-righteousness traceable back to ancient India and the principles of non-violence [6]; an asceticism and corporeal denial long-honored in Christianity [7]; a romanticized community-based model of conformity through shame; a modern grounding in universal human rights discourse; and an increasing portability in the hypervisual, heterotopic global community, hunger strikes span disparate times and places, bridging geographies and epochs as well as motives. Nonetheless, each hunger event is firmly embedded in its own here and now. At the macrolevel it takes place in a specific historico-ideological context, its contemporaneous arena of political conflict and power; yet it plays out in the recessed microcosm of the ravaged, denied body, which is the intersection of the desires and disciplines of the two polarized but dialectic scales of reality. Such collapsed boundaries and distorted dimensions are recurring motifs of starvation warfare, and indeed, the modus operandi of its strategy. With diverse implications, hunger striking thus experiments with the implosion of local-global universes and the blurring/melding of subject-object relations therein and thereof. A challenge to periodicity and geography, the perverse tactic of martyrdom invites chaos into our discrete understanding of categorical certainties and epistemic presuppositions.

Whether resulting in fame or infamy, hunger striking memorializes a stylized performance; one that ritualizes resistance in a silent indictment of authority and a graphic appeal to collective moral outrage. This exceedingly private, excessively social ritual denies much, but desires more: defiling the physical in order to valorize it (perhaps via the will, if not the soul), longing for the fantastical in the awakening of social conscience. It seeks, if not complete revolution, then something more elusive and perhaps more demanding: to raise consciousness and inspire conscience, through a conversion of the moral order.

Yet, what are the significations of this drama? How are power and resistance configured? What do the enterprises of Magritte and Einstein have in common with the hunger strike? The list of questions seems infinite, and reverberating. In this brief essay, I wish to consider the dramaturgical aspects of hunger striking as existential gamble and public spectacle; and explore how the performance of starvation plays with the relativity of both power and corporeality, while postulating an absolute, non-negotiable morality.

PERFORMERS & PERFORMANCE POLITICS

The hunger strike is a keenly methodical act: a deliberately crafted martyrdom that stages interrogation through the simplest tactic of shaming. The curtain rises in descrimina rerum—in the midst of action—and unfolds a story of power and protest which commenced in some amorphous and impenetrable past (in keeping with the obscurity symptomatic of the virtual death of all events that have not yet emerged as 'news', or are no longer 'in the news'). Against this backdrop of asymmetrical conflict and reflexive empowerment the main plot of action is played out: self-righteous martyrdom through private starvation, a paradoxical punishment on disciplinary technologies and their colonizing regimes, a provocative indictment of the dominant system, or the legitimizers thereof.

Opening act presents a non-act: the inverse gesture of rejection, made humbly material in the refusal of food. This mind/body tango of (instinct and will) desire and discipline choreographs a dual chaos, interiorized in the accelerated entropy of starvation; and externalized in the deconstruction of sovereign logic and conventional sociopolitical process. The methodical breaking down of organic order in the body implies a visual, visceral metaphor for the breakdown of moral cohesion in the body politic. Indeed, it may be that the body is the only prop left to the silent inquisitor, a situation several fasters have noted when discussing their chosen tactic of defiance. To quote Diane Wilson, a Texas fisherwoman who undertook a hunger strike to protest petrochemical dumping in the Gulf of Mexico: "I figured my body was the only thing I really had." Vietnam vet Gino Casanova, fasting to call attention to the plight of soldiers missing-in-action, explained it this way: 'I sought a sign, something inescapable.'[8] In these testimonials, we have the body experienced/perceived as the last terrain and ultimate fate of the self; and conceptualized as a potent insignia and medium of determination and moral resolve. Intriguingly with respect to somatic iconography, the starving body conjures an image of nakedness, as if draped in innocence, destitution, vulnerability—the costumes of despair. Stripped of its most elementary accoutrement—the primal accessory of food—the emaciated body needs no uniform, save the simple shroud of humanity, the existential birthday suit of being. Perhaps no dress is more appropriate for the occasion.

With this barren scene the stage is set. From the incipient power of the defiant act to the engagement of participant spectators to the representative authorities indicted for the [presumed] injustice(s) the ensuing performance links actors and audience in a suspenseful drama of discovery and macabre. In this signifying chain, the media play the role of chorus in a Greek tragedy: a facilitator of action—at times sympathetic, at times prophetic—but always essentializing, always the distanced [omnipotent] interpreter. In the contemporary economy of appropriation, however, the striker and media engage in an ambivalent codependency, a mutual exploitation whereby each parasitizes the other to feed its imagery. With every passing day—day 1, day 10, day 24, day 36, 51, 68, 73—suspense mounts, and the plot quickens toward catharsis or drags toward impending tragedy. In this manner, the melodrama both stretches and condenses time, at once accelerating and lengthening existential experience in an affair with spatiotemporal elasticity. Starvation itself is paced by a perpetual present—the insistent physicality of the moment—while the impact of its slow execution relies on the imagination, in anticipating or foreshortening the future.

ACTOR & AUDIENCE

In launching the offensive, the striker seizes the role of protagonist, ransoming the body to declare war on injustice and its worldly proxies. And yet we might ponder the paradoxical agency of self-destruction: is the hunger striker really a protagonist? a centered, autonomous subject (as the United Nations asserts in declaring hunger striking a voluntary act)? It is arguable that this political performance provocatively dramatizes individuality and agency, on the one hand, while conversely hypothesizing an interdependent social contract, on the other. Does not the very effectivity and salience of the hunger strike result from the mutual conscription of the indicted system? One might propose that it is simultaneously by asserting free will and sacrificing it that the hunger striker establishes his/her uneasy empowerment: By projecting agency through the defiance of society qua food, fasting stakes corporeal boundaries as the conclusive arbiter of an autonomous self; while the coercion implied by the defiled body yet posits an organically linked politico-moral order between the individual subject's body and the body politic, wherein social tragedy becomes self tragedy and vice versa. Even the extreme disciplining of self-starvation can be read as both the [decentering] act of a centered subject and intersubjectively determined. This philosophical conundrum is the balancing act of the hunger strike, and sets the stage for its myriad paradoxes.

The mutuality of indictment frames hunger striking as not an act of autonomy, but of entrenched relativity. Inescapably embedded in a sociohistorical and epistemological matrix, the moral punch of the hunger strike derives from a synecdochic paralleling of the striker and the system: The striker assumes the role of tyrant in a metonymical substitution whereby the rebel becomes his or her own oppressor in an act that both confirms the unity of body and mind and disjoins them. Collapsing the roles of perpetrator and victim, the emaciated translates the brutal logic of the system into monstrous action. The chimerical magic continues in the striker's becoming a classic scapegoat, 'standing for' and 'standing in for' the dominant regime. In the manner of the overdetermined tragedy, the hunger strike plays decoy, a dark mirror that reflects, like the secret phantasmagoria of the picture of Dorian Gray, the corruption, decay, and cannibalism of the indicted system.

Reflective and refracted through the prism of representation, the disappearing act of the hunger strike casts the drama of self and other in an ambivalent voyeurism. With shadings of empathy and accusation, fascination and horror, the macabre yet has a mysterious, even seductive quality. The closely patrolled borders of the willful body (or the embodied will?) as the final frontier of resistance are paradoxically founded, not as an opaque sovereignty or an impenetrable fortress, but rather by virtue of a fathomless transparency. For it is only by laying open its anatomy to the cannibalism of the voyeurs' prying gaze that public fasting can hope to accomplish its mission, only by prostituting the tragedy of starvation can it gains its audience. Through fantastical metamorphosis, the hunger strike inverts, perverts, subverts, and introverts, it advertises, externalizes and distorts. The rejection of food as symbolic rejection of the social teases actual and implied boundaries, problematizing the suture between self/other, making slippery the dividing line between social corps and singular corpus, public angst and private suffering. It distorts the connections between surfaces and orifices, exploiting the edges of embodiment, the silhouette of being-in-the-world, being-in-the-body.

ETHICS, ESPRIT & ENERGY

We are left with paradox and the paired suppositions of inside/outside, bounded/fluid, free/determined, death/rebirth, emaciated/emancipated, etc. And yet, in the manner that the velocity of light is taken as an absolute and maximum measure for the evolution of all other energies, hunger striking is staked as an absolute morality, premising that all ethical accounting be measured according to a shared, existential commonality, a commonality foregrounded in the uncomfortable proximity of hunger.

Thus, like the endeavors of Magritte and Einstein, the hunger strike as an act of significance and signification plays with the relativity of taken-for-granted representations, questioning established logic and the way we witness our world. It plays with the dialectic, hinting at the false dichotomy between body-mind, life-art, science-magic, empirical-spiritual. We might find further similarities in the iconoclasm of the masters and the underdog.

For Magritte "critiqued the language of the visual, using literalism to undermine itself" [9]; and Einstein contested the paradigmatic representations of electromagnetic objects in space, having successfully dealt with the representation of gravitational objects in space by substituting his revolutionary equations of relativity. In like manner the mute choreography of hunger striking contests the representational hegemony of language. With its enigmatic antics, centripetal motion (doomed regressive), strange gyrations (around the corps), and dissipated imagery in space, it challenges the inertia of words and attempts to dethrone the privileged canonization of laws to define the codes of ethics, morality, power. By casting off the very premise of exchange in its myriad forms (i.e., the sharing of food, the negotiation of words, the sanctity of laws) hunger striking questions the naturalized values of rhetorical, consumer, and contractual relations, and interjects the body as the ontologically prior and morally superior communicative vehicle. It relativizes power and representation in a way akin to the conceptual revolutions of Einstein and Magritte.

The ritual animates a fantastical polysemy, and establishes the complex synergy and liminality of the act, whereby the striker plays multiple roles: choreographer in a dance of death/rebirth; trickster in the overthrow of representational categories; scapegoat for the sins of society; martyr in a moralizing testimonial; image in the spectator's gaze; blasphemous suicide or saintly altruism; problematic patient or suicidal crusader; practitioner of temper tantrums or stoic integrity?

And so I am left wondering: if Magritte were to paint the hunger strike, how would he represent it? With his signature clouds floating through the receding silhouette of a human figure. If it wrote a calligram would it be a statement on the ethereality of the act? the spiritual, evasive, or absurd? Would he paint a trompe l'oeil continuation of the canvas of the landscape/society spilling onto the canvas of the hunger striker's emaciated body, akin to the mimetic overflowing of his inner canvases, in, for example the Condition Humaine series?

Likewise, how would Einstein calculate the energies of resistance, the velocities of the imagination, the equation of agency? In any event, it is interesting to speculate that the bodily-based, existential morality posited by the hunger striker strives for a universal validity and attempts to transcend the conventional, language-based discourses of power, in much the same manner that Einsteinian physics transcended the earthly particularity of Newtonian mechanics.

Maybe there is an overall Gestalt of being and an absolute measure of humanity.

Kyra Marie Landzelius
Institute for Advanced Studies in Social Anthropology
Pilgatan 19, S-411 22 Goteborg
Sweden

REFERENCES

[1] See: **Beresford, D.**, *Ten Men Dead: The Story of the 1981 Irish Hunger Strike*, New York, Atlantic Montly Press, 1987.
Everett, M., "Veterans Fast For Life", *Christianity and Crisis*, **160**, 1986, pp. 358-360.
Fallon, C., *Soul of Fire*, Iowa: University of Iowa Press, 1985.
"Fasting (Hunger Strike by Palestinian Prisoners)", *The Economist*, Volume bf 336, 8 July, 1995, page 37.
"Hunger Strike Shakes China", *Beijing Review*, **32**, 1989, pp. 22-26.
Kingsford, P., *The Hunger Marchers in Britain 1920-1940*, London: Lawrence & Wishart, 1982.
Norman, C., "Sakharov Hunger Strike Casts Doubt on NAS Plans", *Science*, 24 May 1984.
Sands, B., *One Day in My Life*, Dublin and Cork, Mercier Press, 1983.
"World-Wide: Israel Rushed", *The Wall Street Journal*, 12 Oct. 1992, A1.
Wren, C.S., "Mandela Urges Convicts to End 37-Day Fast." *The New York Times*, 7 June 1991, A7.
[2] "The Farmer Who is Eating Again", *Maclean's*, **28**, 1987, p. 12.
[3] "The Embers of Defiance (Hunger Strike Results in the Dead of M. Astoroff)", *Maclean's*, **19**, 16 Dec. 1988.
[4] **Weinraub, Judith**, "Hungry for the Truth: Jennifer Harbury's Quest for Her Missing Love", *The Washington Post*, 21 Nov. 1994, C1.
"Women to End Guatemala Hunger Strike". *Los Angeles Times*, 21 Nov. 1994, A15.
[5] " 'Peace Pizza' Send to Stallone, Symbolizes Protest of Violence", *Los Angeles Times*, 14 Aug. 1988, II5.
Martinez, Al., "Not Him Again", *Los Angeles Times*, 15 Sep. 1992, B2.
[6] **Sonnleitner, M.W.**, *Gandhian Nonviolence Levels of Satyagraha*, New Delhi, Abhinav Publications, 1985.
[7] **Bynum, C.W.**, *Holy Feast and Holy Fast: The Religious Significance of Food to Medieval Women*, Berkeley, University of California Press, 1987.
[8] "A Viet Goes on Hunger Strike in a Bamboo Cage to Remind Folks of M.I.A.'s (G. Casanova in Kent, Wash.)", *People Weekly*, **24**, 11 Nov. 1985, 137.
[9] **Foucault, M.**, *This is Not a Pipe*, J. Harkness, Trans. and Ed., Berkeley, University of California Press, 1983, Translator's note, page 9.

FENTON F. ROBB

SUPRA-HUMAN INSTITUTIONS AND THE HUMAN CONDITION

INTRODUCTION

In a 'thought experiment' it is postulated that 'institutions', languages, cultures, schools of thought, disciplines and so on, are real-world entities of a natural kind spontaneously emergent from recursive cycles of conversations and communications. Autonomous, autopoietic, and, therefore, self-defining 'living' systems, institutions conform to the 'natural laws' of emergence, energy conversion and entropy production. They emerge from social interaction but not by the deliberate actions of individuals, and they need serve no human purpose. However, they do define what humans can perceive and understand to be real, true, good and causally related. They are not to be confused with man-made, man-controlled contractually-founded fabrications such as organisations, families, corporations, organs of state, and their like. Recognition of institutions as members of the class of natural kind objects and susceptible to conventional scientific investigation would enable a science of 'Institutional Dynamics' to clarify many of the conflicts subsumed under the rubric 'Einstein *meets* Magritte'.

JUSTIFICATION OF APPROACH

The 'thought experiment' is to portray some notions that could become both ideologically and scientifically 'revolutionary' in the Popperian sense. These notions may serve to diagnose the space between Einstein and Magritte; an appreciation of them may point the way towards a greater understanding and improvement of man's condition.

Thought experiments are appropriate when 'normal' science is assailed by unassimilated phenomena. Such experiments serve to question the existing fabric of beliefs that brought a scientific community to its present confusion. According to Kuhn, the data must exist before the experiment begins and the imagined situation must allow the scientist to apply the experimental method in the usual way [1]. That is to say, the experiment has to address some explanatory or predictive hypothesis and it should set out the conditions in which that hypothesis might come to be accepted, or rejected, by some scientific community. The experiment

© 1999 *Kluwer Academic Publishers. Printed in The Netherlands.*

should consider what might constitute, if not a 'proof', then at least some plausible grounds for a tentative belief in the hypothesis, and what might refute it and cause it to be dismissed from further consideration. Kuhn writes, "... on occasions (when conducting thought experiments), even philosophy will become a legitimate scientific tool, which ordinarily it is not". Space precludes the conduct of a full-blown experiment here, but a sketch of some of its salient features will be described.

The agenda for 'Einstein *meets* Magritte' well describes the kind of confusion that such a thought experiment might well address. The confusion is to be seen in the apparent conflict between the scientific and the artistic or relativistic views of the human condition, between the unitary and the pluralist worlds offered respectively by science and the imagination, and, maybe, between the modernist and post-modernist views of man's place in the world.

The insight that arises from the proposed thought experiment should serve, in part at least, to form a bridge between 'natural', 'hard', 'scientific' and 'social', 'soft', 'relativistic' world views and remove some of the confusion which they engender. The result might lead to the shaping, if not the control, of parts of the natural world (things that are here called 'institutions'), the better to serve human purposes.

A 'THOUGHT EXPERIMENT'

The 'data' for the current experiment include recurring reports of systems failures, such as those examined by Perrow and others, the frequently ineffectual attempts to apply science and technology directly to social, institutional, organisational and ecological problems, the decay of notions of generalised social justice in the face of an uncontainable efflorescence of hedonistic individualistic consumerism [2]. More particularly, the data include growing feelings of helplessness, isolation and alienation, frustration and disorientation and the paranoid delusions and flight to superstitious beliefs and practices that beset increasing numbers of individuals as they address their own everyday psychological, social, technological, political and economic problems.

At the extreme, the data include the patent hopelessness of what Cornel West has called 'walking nihilism', not simply a confusion of ideas, but the passive living-out of meaninglessness in a haze of dope, fear, and violence against the Other and against the Self [3].

The experiment entails abandoning the anthropocentric tendencies of both 'normal' science and relativism. These tendencies have given rise to much of the confusion and feelings of helplessness (the topics of this Conference) by attributing to individual people and to man-made organisations the responsibility and the power to understand and alter all the social processes, and many of the physical ones too, that affect the human condition. This anthropocentric view was summed up in the infamous phrase of ex-Prime Minister Thatcher: "there is no such thing as Society, there are only individuals".

The attribution of all responsibility for the human condition solely to the activities of individuals and congregations of individuals, rather than to the operation of forces of nature, blind chance, divine intervention or other supra-human influences, has created cultures of blame and envy and caused spasms of collective rage and hatred against particular people, classes of people and organisations of people. Blaming such scapegoats excuses making any further examination of the possible causes of such decayed human conditions. In consequence, the so-styled 'social sciences' have degenerated into studies, not of the problems of sick societies, but of the problems of socially sickened people. Because there has been no substantial model of the institutions of society, little or no scientific attention has been directed to social issues as such.

It is held here that the anthropocentric assumption (that deliberate human action, singular or collective, lies at the heart of all human problems) has diverted the attention of science away from a whole group (a 'Kingdom' perhaps) of living entities, barely recognised hitherto, that are having an increasingly profound influence on the human condition. These entities may have many properties in common with the 'social' systems spontaneously emergent from interactions between individuals in other species such as swarms, herds, hives, schools, skeins and so on—but such considerations are beyond the scope of this short paper.

CONVERSATIONS AND COMMUNICATIONS

For the purposes of this experiment, it is held that human interactions, in the form of 'conversations' and 'communications', are real experienced events. This assumption follows Hume's [4] assertion that beliefs in matters of fact are ultimately founded on some experience or memory of experience, or else they are without foundation. The contrary position (the denial that human interactions are not experienced as real world phenomena) seems to lead to the kind of 'false philosophy' that Hume would enthusiastically 'consign to the flames' [5].

The interactions of people (conversations and communications) are here taken to be homologous with the interactions of physical systems in nature. Energy is transformed and entropy is produced. Also, as predicated by systems understandings, such interactions may, under certain conditions, give rise spontaneously to entities of a higher order. What emerge from human interactions are here called 'institutions'.

'Conversations' are mediated by people talking to each other, and to themselves. Conversations entail interaction between individuals, or reflection by one individual, such that 'learning' takes place. Learning involves the making, shaping and reshaping of the cognitive systems of individuals. The prior conditions for conversation are that there be agreement to converse, a minimal common language and some commonality of cognition available to the individuals involved. There must also be different ignorances or dissonances to which conversations can be addressed. Conversations are about discriminating objects from amorphous grounds, and signing them. They are about making and revising distinctions and definitions, forming and modifying categories, and attaching to these categories social, political, economic, aesthetic and moral evaluations. Conversations indicate what phenomena, objects and events do exist in the world and what abstractions can be drawn from it, which of these should be addressed as being of objects of interest in particular conditions and which should be ignored. Conversations enable many and various objects and events to be recognised as having similarities to be described and evaluated in generic terms. Conversations define what existential, temporal and causal relations are recognisable and plausible, and what are not. Conversations form hypotheses, and, after 'communications', such as those described below, have taken place, they form theories about the world held to be plausible, valid or truthful by groups of people in society.

The processes involved in conversations are reiterative and cyclic [6]. Conversations not only create and alter the cognitive structures of those involved; they also bind people who share the same cognitions closer together into social groupings. By the same tokens, they also serve to separate people with different world views from each other, disabling conversations and communication and thus substantiating the Other.

Important for this experiment, conversations are observed to generate prescriptions for human action and to specify what kinds of action are required of those who share in a particular system of belief. For example, the many and various criteria that define what is plausible, good and true in the different scientific disciplines, in religions, cultures and art forms have been, and are being, generated and shaped by conversations. Conversations also address the human activities, individual and social, that are defined as 'norms' and that are to be tolerated and promoted by the groups of people bound by common understandings of objects, events, categories, values and so on. They also define the abnormal, what is undesirable and what should be excised. All the various Einsteinian and the Magrittian views of reality are so shaped.

The outcomes of conversations are predictions about the future states of affairs, and these outcomes become the objects of 'testing' by communicative actions. This relation between conversation and communication forms part of the self-productive cycle of institutions.

Conversations end when there is total agreement between the individuals, when there is a breakdown of the agreement to converse or when the language being employed is found incapable of conveying what is there to be learnt [7]. The ending of a conversation may signal the ending of the emanation of an institution.

'Communications' initiate the human actions and interactions that are constrained and shaped by conversations. Typically, they consist of instructions, prescriptions, orders, advice, rules and programs for action. Communications are possible when the participants share, or appear to share, common conceptual structures in respect of the language used for the communications. Communications may not necessitate learning or the alteration of the cognitive structures of those involved, but they do enable the 'testing' of the anticipated outcomes of situations that have been the topics of conversation. The criteria for testing are laid down by conversations, but testing is a distinct communicative activity directed to endorsing or refuting the plausibility of the cognitive structures created by prior conversations.

The actions resulting from communications expose disagreements, dissonances, inconsistencies and contradictions. If the outcomes of prescribed actions are not quite like those anticipated by the preceding conversations, the differences constitute feedback that may generate further conversations and further revisions of the socially negotiated cognitive structures. The relations between communications and actions, like those within conversations, are cyclic as successive prescriptions are applied to situations, either in the effort to bring about the outcomes that were anticipated by the preceding conversations, or to initiate further conversations in order to reduce the dissonances between the concepts held in view and the experiences of the world [8].

In practice, conversations and communications are deeply intertwined and difficult to distinguish; but what is of theoretical interest is their generic relationship. Together, conversations and communications constitute recursive cycles of mutual causal processes [9]. And it is these that constitute the relations of production of the autopoietic and autonomous supra-human entities here called 'institutions'.

The relations between conversations and communications (not people) are the 'components' that produce and sustain the spontaneous emergence of institutions. Members of the scientific community will recognise the foregoing description of the emergence of institutions as a generalised description of progressive scientific programmes and of the emergence of novel scientific paradigms [10].

Institutions should not be confused with man-made social structures such as firms, trans-national corporations, families, societies, clubs, the organs of state and their like. In this paper these are called 'organisations'. 'Organisations' are different from institutions. They are constructions created by implicit or explicit contracts between humans, involving role-playing, the sharing and delegation of authority and responsibility, the division of labour, specialisation, commitment and participation. Organisations can be defined by their creators in terms of these contractual relations, but they can also be defined, with equal validity, by observers with different interests; for example, by members, clients, customers, owners, managers, citizens, people in different demographic categories and so on. The boundaries of organisations defined by different people may vary widely depending on the reasons for drawing them. But no distinction can claim any special validity. Because there can be none but arbitrary grounds for agreement about their boundaries, and about what they contain and what they do not contain, generalisations (other than this one!) cannot be made about organisations. Organisations, therefore, cannot be objects for scientific study (despite the claims made by management theorists). Organisations belong to a relativistic Magrittian world. The evanescent and multi-faceted nature of organisation may be why, despite the expenditure of great resources on organisational research, nothing approaching a science of management has achieved credibility as yet.

In contrast, 'institutions', being autonomous and autopoietic, irreducible and primitive of their kind, define themselves. The existence of institutions, unlike that of organisations, depends not at all upon the predilections, intentions or interests of any observers, but upon the necessity of certain relations of production required to allow their continued regeneration, adaptation, and existence over time. The boundaries drawn by observers cannot, except by chance, correspond with those of any institution. The institution's definition of itself and the boundaries around those relations required for its self-production at any instant, depends on no observer. In a similar way only the institution itself can define which components (conversations and communications) and what relations between them are necessitated for its survival. Institutions can survive only if their components and the processual relations between their components are changing constantly.

It is argued that institutions are real-world things-in-themselves, that they are proper objects for scientific study and that, given such study, useful generalisations could be made about the conditions of their existence, their properties and the effects that they have on the human condition. Institutions belong to a natural Einsteinian world.

INSTITUTIONS

So, in this thought experiment, it is advanced that the seemingly chaotic interactions of human conversations and communications promote the spontaneous emergence of ordered entities, 'institutions'. These are typified as cultures, languages and schools of thought, and what Hume called 'customs and habits'. These, he contended, embodied 'the ultimate principal of all conclusions arising from human experience'.

Just like other emergent entities, institutions have properties quite different from those of the processes from which they emerged. Emergence entails morphogenetic change, the exact nature of which cannot be predicted. This is because very small differences in initial conditions, and in the conditions present at the point of emergence, the bifurcation point, can cause very large differences in the properties of the emergent entity [11].

So, the experiment suggests, institutions are quite different in kind from the cycles of conversations, communications and actions from which they emerge. They are of a 'higher' order than that of their components, and they define their own boundaries through the enactment of these conversations and communications. Institutions are autonomous, and self-defining, 'autopoietic', 'living' systems [12]. Thus, it is no fallacy to say that they are 'purposeful actors' engaged in preserving their boundaries and in regenerating their own components (their conversations and communications). Indeed, if it is wished to characterise their 'purpose', it is sufficient that they preserve the processes of their self-production. But it would be fallacious to anthropomorphise them and attribute human-like properties to their animate processes.

The central criticism of social holism of the kind similar to that proposed here, has been that it hypostatizes and attributes real existence to what are really abstractions. Examples include the family, the state, the firm, the club and the organisation as it is seen here. This criticism is justly levelled against notions that attribute real existence and even life-like properties to organisations, as in Living Systems Theory [13] and the theory of autopoiesis as interpreted by Zeleny and Hufford [14]. However, if 'institutions' are indeed natural kind entities, then such a criticism cannot be levied against the theory of institutions as expressed in this thought experiment.

Although institutions define what criteria shall be applied by their entrained humans to determine what is real and unreal, true and false, cause and effect, and what is desirable or undesirable in their domain of interest, they *need* serve no purpose other than the maintenance of their own processes of self-production. These criteria are, as it were, allopoietic by-products, even waste products, of the processes of institutional self-production.

The energy required to sustain the self-productive processes of institutions has been, to date, that derived from the human lifetime devoted to the continuation of the relevant conversations and communications. Because of this, institutions appear to serve at least some human purposes or, at least, the purposes of some humans, and they seem to be created by deliberate human effort. Humans appear to be the creators of institutions and institutions seem to be the objects of that creation. In this experiment, these roles are quite reversed. Human 'individuals' are people defined in institutional terms, and they owe their existence and their specific identity to the institutions that define them and attribute to them properties, relationships in their groupings and 'social rights' [15]. The notion of the human individual is of quite recent origin and it might not be inconsistent to attribute the emergence of that notion to the spread of literacy and formal education, for both of these are likely to be effective catalysts for the emergence of institutions that create identity by accelerating conversations and communications.

It is the mistaken conflation of social artefacts (organisations) with spontaneously emergent entities (institutions) that has been the cause of much confusion and human unhappiness, of the many cultures of blame and of inflated notions of individual human responsibility.

Institutions provide the tests for the admission and rejection of the distinction of objects and events from their amorphous grounds and of structures of belief (and, sometimes, of believers and non-believers) to their domains. Institutions admit not only the distinctions between facts and illusions, but also the descriptions of worthy and unworthy objectives and actions. Most importantly, institutions determine what constitute verification and falsification of their own metaphors or paradigms. But these features of institutions are neither necessary nor sufficient for their emergence, nor for their continued existence. There are probably many other ways in which they can ensure their own continuation. Institutions have a unifying effect on the mental constructs they promote, each within its own conceptual domain. But each depends on preserving its boundaries, of distinguishing its own domain, from those of its neighbours in the general field of related domains. Institutions appear to 'compete' with each other for human lifetime and they decay if it is denied to them.

Institutions also seem able to interact with each other in complex ways to enable the emergence of hierarchies of even higher-order institutions of institutions. The experiment suggests that a new vocabulary, perhaps a new calculus, is required to describe the many levels and the many trans-level interactions that occur in the institutional world.

Institutions are realised as virtual bodies of received wisdom and knowledge, of facts and of understandings about existential, causal, moral and temporal relationships. Each defines its own ontology, methodology and epistemology. It may be that the phyla (or divisions) of the Kingdom of institutions would distinguish institutions in terms of the relationships between their conversations and communications; say, differentiating between those that try to preserve a unitary conversation but are greatly interested in the feedback aspects of those relations ('scientific' institutions), and those more interested in generating variety in their conversations, with less interest in testing through communicative activity ('relativist' institutions).

This experiment has been conducted within the constraints of both of the main kinds of the contemporary institution under consideration. In as much as the aim of this paper is to generate a unitary view of the natural and the physical worlds, the approach is 'scientific'. To the extent that the experiment postulates a new order of being, the approach is 'imaginatively Magrittian'. As to the medium of its conversation, this paper probably fails to satisfy the plausibility criteria of either approach.

The crucial test of the falsifiability of the hypothesis that institutions are supra-human real world things of a natural kind has yet to be considered. The falsification of this hypothesis takes the form of the Epimenidian paradox ('All Cretans are liars' said Epimenides, the Cretan). The boundaries of an institution can be defined only by that institution. All institutions are of a higher logical order than the conversations and communications that we humans require to describe them. The hypothesis that institutions are supra-human objects will be falsified if and only if we can observe the boundaries of any institution directly and completely and if we can fully describe them.

TECHNOLOGY AND INSTITUTIONS

It would seem that, in the past, some 'critical mass', as it were, of conversation and communication was needed before any institutions could emerge and survive successfully. Maybe this was because these processes could be conducted only at the pace of human speech. Now, the written word, the typewriter and information technology, the mass media and interactive computer networks, enable conversations and communications to proceed very rapidly indeed.

All these conversations are inventing new categories and subcategories of people and generating new systems of belief and new social values at a high rate indeed. These speedy interactions between 'virtual people' may be allowing the emergence of large numbers of very small institutions, some of which may threaten or undermine the power bases of their larger predecessors. These newly emergent institutions are being energised not only by the lifetime of the participants in contemporary conversations and communications, but also by the human energy potential stored in the vast data bases which are now spontaneously connecting themselves together as interactive networks.

In the recent past, it would seem that there were relatively few institutions at any place and time and thus only a few criteria of truth, plausibility, and sensibility. Now, there are very many institutions, connected in many complex and discrete ways. So many are the criteria available that almost any notion can be found an institutional 'home', a system of rationality, or a-rationality, to justify it. At present the choice for the human person seems to be between taking the Humean position and being resigned to scepticism about an external world, this is the strictly scientific view; or submitting to some arbitrarily selected institutionalised truth system emerging from the Tower of Babel, the Magrittian position. The realisation that institutions are creating and commanding both 'worlds', though it throws little light on the central philosophical issues, should, at least, emancipate us from having to make such a choice.

If the emergence of institutions can be energised by relatively small domains of conversations, by non-human activity, or, more strictly, by aggregated human lifetime (say, that stored in data banks) the views of the sources of institutions, expressed above, will have to be enlarged. The technology of surveillance, just one aspect of the contemporary technological revolution, serves to illustrate this point.

The technology of surveillance plays an important role in stabilising powerful institutions and it spreads now into the most intimate aspects of human lifetime. Each surveillance system specifies certain aspects of the individual that some institution requires to be monitored in order to promote its own survival. People are thus defined and redefined as unique individuals by institutions that, at the same time sort them into physical, psychic, economic, social and moral categories.

Foucault has pointed out that the individual is becoming more individuated by the processes of surveillance; what it is to be an individual is being more precisely defined by institutions employing the technologies of surveillance [16] [17]. The individuality of the originators of the surveillance is becoming more obscure; the educators, police, judges, professionals, experts and other guardians of the 'good' are only the faceless agents of ever more remote, amorphous and autonomous institutions that make the need for surveillance seem rational. Soon the development of technology may remove humans from such activities altogether. Yet the technology seems to amplify the myth that individuals, and not suprahuman institutions, are responsible for their own destinies. Institutionalised thought and its associated technology are making institutions ever more opaque.

How the electronic mediation of conversations and communications and the advancing technology of surveillance will affect the emergence of institutions remains to be seen, but it is already apparent that the perpetual memories and eternal records that are being created now have the potential to hold to account people and social organisations that might hitherto have escaped scrutiny [18].

TOWARDS A SCIENCE OF 'INSTITUTIONAL DYNAMICS'?

Because institutions are of a higher order than human conversations, it is concluded that, although the class of entity called 'institution' may be recognised by us, individual institutions cannot be defined exactly, nor can their internal processes of self-production be observed by us directly. If humans cannot observe the workings of the particular processes of self-production of specific institutions, they are not in a position to affect them.

If, as postulated here, institutions do conform to the same 'natural laws' (of emergence, energy conversion and entropy production) as other real-world systems, if they are entities of a natural kind and, therefore, irreducible of their kind, then observation of their components, their conversations and communications, will shed little or no light on the dynamics of their self-production [19]. Other means must be found. In the natural world, entities of different orders, say, from galaxies to single-celled organisms, are rarely experienced directly but require specifically different modes and means of observation. So, too, is it the case that when the existence of things of a new order is suspected, new instruments of observation have to be fashioned with which to observe them. So new instruments of observation are required in order to inquire into the factors affecting the emergence and survival of institutions. So how institutions emerge and define themselves, how they utilise human resources and how they interact with each other appear to be proper objects for study by scientists of 'institutional dynamics'. Humans may never be able to intervene in the emergence of institutions or direct their processes, but they may be able to devise instruments to infer whether particular institutions are emergent or in a state of decline and, maybe, to anticipate how they may affect the human condition.

By appreciating the dynamics of the processes of formation and maintenance of institutions as a class of real world objects, we may be able to shape them in some way, the better to serve human purposes. But first, institutions must be distinguished clearly from man-made, potentially controllable organisations, and recognised as 'living' entities of a natural kind and of a higher order than human conversations.

This thought experiment enables an imaginative leap to be taken. Many things that affect human affairs can be seen as no different in kind from other phenomena in the natural world. When human affairs appear confusing and beyond our control, instead of blaming particular individuals or factions, chance or some supernatural cause, we may do better if we turn our attention to how institutions emerge to affect our condition and what the general laws of institutional dynamics might be.

Fenton F. Robb
Ocean Eye, Eyemouth,
Scotland, TD14 5ES UK
e-mail: robb@dial.pipex.com

REFERENCES

[1] **Kuhn, T.S.**, "A function for thought experiments", In: **Hacking, I.** (ed.), *Scientific Revolutions*, Oxford University Press, Oxford, 1981, pp. 6-27.

[2] **Perrow, C.**, *Normal Accidents: Living with High-Risk Technologies*, Basic Books, New York, 1984.

[3] **West, C.**, "Anders Stephanson interviews Cornel West", In: **Ross, A.** (ed.), *Universal Abandon? The Politics of Postmodernism*, Edinburgh University Press, Edinburgh, 1989.

[4] **Hume, D.**, *An Enquiry concerning Human Understanding*, In: **Selby-Bigge, L.A.** (ed.), revised Nidditch, P.H., 3rd. edition, 1975, Clarendon Press, Oxford, p. 46.

[5] **Hume, D.**, ibid., p. 165.

[6] **Pask, G.**, "A conversation theoretic approach to social systems", In: **Geyer, R.F. and van der Zouwen, J.** (eds.), *Sociocybernetics*, Martinus Nijhoff, Leiden, **Vol. 1**, 1978, pp. 14-26.

[7] **Pask, G.**, "The Limits of Togetherness", In: **Lavington, S.H.** (ed.), *Information Processing*, Amsterdam, North-Holland, 1980, pp. 999-1012.

[8] **Festinger, L.**, *A Theory of Cognitive Dissonance*, Row Peterson, London & Stanford, 1962.

[9] **Maruyama, M.**, "The second cybernetics: Deviation-amplifying mutual causal processes", *Scientific American*, **51**, 1963, pp. 164-79 and 250-80.

[10] **Lakatos, I.**, "History of Science and its rational reconstructions", In: **Hacking, I.** (ed.), *Scientific Revolutions*, Oxford University Press, Oxford, 1981.

[11] **Prigogine, I. and Stengers, I.**, *Order out of Chaos: Man's New Dialogue with Nature*, Heinemann, London, 1984.

[12] **Maturana, H.R. and Varela, F.J.**, *Autopoiesis and Cognition: The Realization of the Living*, D. Reidel, Dordrecht, Holland, 1980.

[13] **Miller, J.G.**, *Living Systems*, McGraw Hill, 1978.

[14] **Zeleny, M. and Hufford, K.D.**, "The application of autopoiesis in systems analysis: Are autopoietic systems also social systems?", In: **Fleischaker G.R.** (ed.), *Autopoiesis in Systems Analysis: A Debate'. Special Forum Issue, International Journal of General Systems*, **21**, 2, 1992.

[15] **Douglas, M.**, *How Institutions Think*, Routledge and Kegan Paul, London, 1987.

[16] **Foucault, M.**, *The Order of Things Pantheon*, New York, 1970.

[17] **Foucault, M.**, *Power/Knowledge; Selected interviews and other writings 1972-1977*, **Gordon, C.** (ed.), Brighton, The Harvester Press, 1980.

[18] **Robb, F.F.**, "In defence of conversation: a polemic against information technology and an agenda for its future", *Systems and Cybernetics in Management*, **19**, 1/2, 1990, pp. 15-25.

[19] **Robb F.F.**, "On the application of the theory of emergence and of the law of maximum entropy production: foundations for a theory of general evolution", *Systems Practice*, **3**, 4, 1990, pp. 389-99.

CZESŁAW MESJASZ

STABILITY, TURBULENCE, CHAOS?: SYSTEMS ANALOGIES AND METAPHORS, AND CHANGE IN CONTEMPORARY WORLD POLITICS

Stability, that much overburdened word with unstabilized definition

Richard Bellman [2a]

Stability—political and economic conditions that satisfy our interests

From the "Doublespeak Dictionary for the 1990s"
Edward S. Herman [21]

INTRODUCTION

The early cybernetic ideas suitable in describing the bipolar world do not seem very useful in exploring changes in the contemporary international system. Modern 'systems thinking' has, however, yielded other concepts. Chaos theory, non-equilibrium thermodynamics, catastrophe theory, autopoiesis and self-organization theory, synergetics, etc., appear to be useful for international studies not as sources of mathematical models but mainly as sources of analogies and metaphors. It is connected with increasing interest in applications of 'post'-structural, -modernist, -rationalist, approaches in international relations studies. The main aim of the paper is to examine what are the advantages and disadvantages of applications of analogies and metaphors based on systems thinking in describing and explaining change and *status quo* in the post Cold-War international system. As it is always the case for interdisciplinary research, in international relations studies, and generally in social sciences, numerous errors are made in applying those new terms. Therefore it is necessary to present the ways how analogies and metaphors associated with 'systems language' can support studies of dynamics of international systems, or generally speaking, social systems.

Inspiration for the paper has come from the coincidence of two phenomena from *praxis* and two from theory of international relations yet pertinent also to other social sciences.

© 1999 *Kluwer Academic Publishers. Printed in The Netherlands.*

- Unpredicted and overall peaceful termination of the Cold War and of the bipolar world.
- Emergence of opinions about achieving a certain 'mature' or even ultimate stage of human development—'The End of History' (even if only in a Hegelian sense), characterized by liberal-democratic political systems and free market principles. In a more 'technocratic' and 'systemic' terminology this stage is characterized by processes of growing variety, differentiation, and self-organization at various levels of hierarchy of social systems.
- Awareness of limitations of applications of mathematical modelling in international relations studies and in associated areas.
- Increasing interest in applications of 'post'-structural, -modernist, -rationalist approaches in international relations studies, cf. [12, 43].

The main aim of the paper is to examine what are the advantages and disadvantages of applications of analogies and metaphors based on systems thinking in describing and explaining change and *status quo* in international systems. Special attention is paid to their normative interpretations. We intend to present an introduction to the studies allowing to answer what is the impact of systems analogies and metaphors used in international relations studies on dynamics of broadly defined international systems (social systems). What place are they taking in the self-referential and autopoietic development of those systems?

The concept of 'stability' can serve as best example for demonstrating the impact of 'classic', cybernetics-based systems thinking on the language of international relations studies (peace research, security studies). At present, the term stability is becoming less useful in studies of international systems. The notion itself, however, has gained a well-established interpretation in the language of practice.

In theoretical discussions it is being replaced with other analogies and metaphors borrowed from systems thinking. Turbulence and chaos seem to be most representative. Their applications seem particularly germane in portraying emerging post-Cold War international situation. However, the above notions are frequently confused by simplifications and/or too far-reaching conclusions drawn from descriptions and explanations based on systems language.

SYSTEMS MODELS, ANALOGIES, AND METAPHORS IN INTERNATIONAL RELATIONS STUDIES

It is not possible to produce a commonly approved definition of systems theory, *systems thinking*, systems approach, etc. [34]. The shortest explanation is that systems thinking can be understood as a specific, holistic approach aiming at identification of universal laws governing collective phenomena in various entities (physical, biological, technical, social). Since the very beginning, advocates of systems thinking would tend to represent a new paradigm or, at least theory—General Systems Theory, connected with more specific areas like systems modelling, systems engineering, or systems analysis. There are various interpretations of mutual relations between cybernetics and systems thinking yet for the sake of brevity we can agree with von Bertalanffy that the former can be regarded as a part of the latter [6].

International relations studies, security studies and peace research have become one of the most attractive areas for application of systems models, either taking shape of sophisticated mathematical formulae, and/or more or less advanced computer simulation models. The climax of enthusiasm for such modelling was observed in the 1970s with growing disenchantment and less optimistic expectations getting ground in the same period, cf. [19] [32]. New simulation models are still developed, e.g., [9], yet today they are treated mainly as educational tools without excessive expectations for their predictive or prescriptive capacity.

There exists another barrier impeding usefulness of any rigorous modelling of international relations. Significance of models is always measured by their impact on social *praxis*. Models are evidently useful in theory and/or in policy making, although their deficiencies make validity of their use rather limited. Sometimes publicized results of modelling can influence society, e.g., the first two Reports of the Club of Rome [31] [33]. It is, however, mainly the everyday language of politicians and journalists along with more or less precise 'scientific' utterances that both describe and determine international reality.

Systems analogies and metaphors acquire a specific *normative* sense while used in international relations theory and practice. Due to their origins in 'rationalist' disciplines - mathematics, physics, chemistry and biology they are treated as *objective* and *scientific* in a rationalist sense. Thus their applications, in addition to enhanced explanative validity, *by definition* obtain supplemental, 'sound', normative—predictive and prescriptive, legitimacy in any debate on international relations and on other social issues. Consequently, in those applications, but not only, their metaphoric sense is neglected or misinterpreted.

Analysis of the specific value-ladening can be illustrated with analogies and metaphors associated with the notions: stability, turbulence and chaos. In that case systems thinking can be seen as collection of 'image producing fields' and international relation studies (social sciences) as a set of 'image receiving fields'.

THE CONCEPTS OF STABILITY, TURBULENCE AND CHAOS IN SYSTEMS THINKING

Stability

The origins of discussion on stability in cybernetics can be traced in the works of Richard Bellman whose concepts, developing the ideas of Lyapunov and Poincaré proved applicable in mathematical modelling of automatic control systems [5]. In cybernetics stability is regarded as a positive state although with some exceptions [2b].

In the traditional sense, close to the original ideas of mathematics, although mostly appealing to intuition rather than mathematical rigour, stability means that if a system is perturbed slightly from its equilibrium state, all subsequent motions remain in a correspondingly small neighbourhood of the equilibrium [14]. Equilibrium may be defined as the steady state of a system or the state that could endure under disturbance, or the state to which the system (if stable) returns despite disruption.

Stability can also be identified as probability that: 'The quality of the system state is preserved during the given time against the disturbance with a certain probability' [46].

All the above definitions, epitomising the state-of-the-art of knowledge in the 1950s and in 1960s, were relevant to cybernetic systems, based on feedback control. Cybernetic systems can also be ultrastable, i.e., be able to exhibit step-wise adaptation to its environment until equilibrium is reached. In other words, a system is ultrastable if it is able to respond not only in a predetermined manner but also by changing its programme of responses when a programme fails. Stability and ultrastability are associated with the system's ability to keep and restore the values of its parameters within certain limits. This phenomenon is called homeostasis [2].

Turbulence and Chaos

Turbulence, a term widely used in physics (hydrodynamics), has been introduced into the studies of world politics by James Rosenau [42]. Rosenau refers to extended understanding of turbulence including concepts relating to non-equilibrium thermodynamics elaborated by Ilya Prigogine and extensively applied in social sciences as source of more or less carefully elaborated analogies and metaphors.

Literature on applications of non-equilibrium thermodynamics, and chaos theory in social sciences is abundant. Considering it as a group of 'image producing fields' for various analogies and metaphors, the following features of all those concepts are to be exposed.

Non-equilibrium thermodynamics or dissipative structures theory, provides a new pattern of change [39]. According to that theory new structures emerge not in changing the state of equilibrium but in the state far from equilibrium. It is the fluctuations that enable a system to move from one state to another. Evolution of such systems can be depicted as follows.

Before the formation of a new structure, fluctuation is the trigger for the self-organization of systems. Close to the point of change—'critical point', 'threshold point'—the system is extremely unstable and unpredictable due to *bifurcations* that determine the path of system's evolution. In addition to those concepts non-equilibrium thermodynamics enriched the language of science with such utterances like 'From Being to Becoming' [38] and 'Order out of Chaos' [39].

Chaos, another notion stemming from systems thinking and popularized among laymen by Gleick [17], remains currently one of the most 'trendy' terms of social sciences. In mathematics and mechanics the term 'chaos' is assigned to apparently random or unpredictable behaviour in systems governed by deterministic laws [30] [17]. A more accurate term 'deterministic chaos' which is used in those areas, at first glance looks antonymous because it connects two notions commonly regarded as incompatible.

Chaos theory opens new intuitive comprehension of randomness and probability. Traditionally, it was commonly believed that the world is unpredictable because it is complex, composed of myriads of elements standing in interactions. Chaos theory shows that systems may behave unpredictably despite their seeming simplicity and the fact that the forces involved are governed by well-understood physical laws. The common feature of those systems is a very high degree of sensitivity to initial conditions and to the way they are set in motion. Chaos theory provides also a new understanding of systems change, enlarging the collection of analogies and metaphors applicable in social sciences [28].

Study of change can be put in the historical three-stage framework illustrated with the development of science [39]. The earliest theories focused primarily on the so-called steady or equilibrium states.

The second stage began with recognition of periodic fluctuation (oscillation). The third type of change is highly unpredictable although it stems from deterministic considerations. It is ruled by a '*strange attractor*' represented with a fractal, a complex geometric object that reveals more details as it is magnified, and one that contains the blueprint for its reproduction in each fragment (self-similarity).

What seems essential in the third pattern of change is its unpredictability resulting from *sensibility to change* in initial conditions. It is usually depicted by the famous analogy applied for the first time in meteorology, called 'Butterfly Effect' suggesting that the mere flapping of a butterfly wing can change the weather [27]. Such unpredictability undermines the value of any modelling of complex phenomena, especially in social sciences and imminently forces us to treat all those models merely as sources of analogies and metaphors.

The above, by necessity superficial, survey, illustrates only a part of the richness of discussion on change of systems. Trying to summarize the above considerations as to make the 'image-creating fields' more comprehensible we may state that all the concepts relating to systems stability and change can be divided into two groups:

- Ideas explaining maintenance, regulation and control, equilibrium-like behaviour, and some aspects of adaptability of already existing systems; they are primarily connected with early stages of cybernetics—automatic control systems, adaptive systems.
- Ideas explaining emergence and evolution of structurally new systems—'non-equilibrium systems', 'living systems'.

The most important contrast between the two groups is that cybernetic, or regulation- and control-concepts, are deterministic (even if disturbances are stochastic), and deal with restoration of stability when the system is destabilized. Concepts of evolution of the equilibrium state and system reconfiguration may be stochastic/deterministic and deal with instability as an opportunity for structural change.

STABILITY, TURBULENCE, CHAOS AND THE LANGUAGE OF INTERNATIONAL RELATIONS STUDIES

Stability

International relations theory and practice can be viewed as a set of 'image-receiving' fields in our discussion on systems analogies and metaphors. Bearing in mind the variety of definitions of systems presented in writings in international theory, for the sake of brevity, we may assume that our discussion pertains to the systems composed of *nations-states* along with any other units that may directly or indirectly influence changes in international systems (world system), e.g., international organizations, regional organizations, social movements, individuals.

Several surveys of definitions of stability in international relations already have been published, e.g., [16] [7] [36] [46]. The term 'stability' was brought into international relations studies by Lewis Richardson

[40]. The concepts of stability and equilibrium of international systems (or in international relations), were thoroughly developed by Morton Kaplan [23] [24], Karl W. Deutsch and J. David Singer [13], Kenneth Waltz [50], George Liska [26], and Richard Rosecrance [41].

They tried to use cybernetics in depicting dynamics of international systems and emphasized that those systems are not simply stable but in Ashby's sense ultrastable [24].

The cybernetic pattern of equilibrium and stability lay at the origins of the discussion 'polarity vs. stability' with Deutsch and Singer [13] favouring multipolarity, and Waltz [50] favouring bipolarity as necessary for stability of international systems.

The pioneers of applications of the concept of stability in international relations tried to use their contemporary cybernetics in describing dynamics of international systems. However, they did not find any adherents. Most authors narrowed the meaning of stability to absence of war. Such a narrowing had several meaningful consequences.

In the language of systems thinking stability is treated a *dormant metaphor* corresponding to the Latin *stabilitas* that means ability to stand firmly or the state of being stationary [37a]. This quasi-literal sense of stability proves sufficient in verbal delineation of mathematical definitions.

Narrowing the definition of stability meant absolute change in its metaphorical character. It was transformed from dormant (quasi-literal) metaphor into a dead (frozen) metaphor with its own literal meaning. Using metaphorical language we may state that stability was 'killed' and lost its heuristic value as a dormant metaphor.

In that stage stability became something that might be called a homologue of 'perpetual peace' of a technocratic era. The word 'peace' seemed emotionally loaded, irrelevant to modern age. Likewise the concepts of 'positive' and 'negative' peace motivated some scientists to use another, more precise term. Maybe the abuses of the word 'peace' in the Soviet anti-Western propaganda were also one of the reasons for search for another term depicting the absence of war and other major conflicts.

Nevertheless, since the works of the 'classics' of international relations studies, stability in its confined sense has also permeated into the language of practice of international relations—policy making, diplomacy and journalism. Absence of military conflict as a positive situation is indisputable. In some instances, however, an emotionally neutral term can be used for concealing the interests of those depicting situation as stable, cf. [21].

Such an approach had an important theoretical repercussion. Change of metaphorical interpretation of stability tied international relations studies with cybernetic system analogy but in a very simplifying manner. Applications of stability in a narrowed sense were caused by a hidden assumption of linearity, associated with predictability (return to equilibrium state after disturbance). On the contrary, the original meaning of stability of international systems, introduced by Deutsch, Singer and Kaplan, derived from ultrastability, had a direct link to nonlinear explication with little chances for predictability.

That connection led to significant *weakening* of methodology of international relations studies by compelling it to notorious efforts to operationalize mathematical models of intricate multidimensional international crises.

In social sciences, especially in organization and management theory, evolution went in the opposite direction. After intensive mathematization reflected in 'systems analysis' and 'systems engineering', limited usefulness of classic cybernetic analogies and rigorous models was recognized in the 1970s, cf. [4] [25]. On the contrary, international relations studies have been too long preoccupied with such problems as for instance conflict and structure.

Without delving too deeply into the grand debate stability vs. polarity, and referring to the analogy with classic cybernetic system, the question of stability and polarity could be put in the following terms: What structure of international systems treated as a cybernetic system is more prone to a certain kind of disturbances, in that case war or major crises?

In automatic control theory such a question can never be answered univocally without additional assumptions usually simplifying the problem, making it 'linear' or at least resolvable numerically. It is not possible to achieve this for social (international) systems. Even recent, more advanced definitions of factors influencing stability and polarity, e.g., [7], still fall short from required exactness. Empirical evidence based on those or similar assumptions can lead to completely opposite inferences, creating a 'bipolarity' among scholars [26].

There are two reasons that caused such a prolonged utilization of classic cybernetic system analogy along with associated metaphors in international relations theory.

First, it was the *bipolar world*, no matter how refined in theory [7], that solicited that analogy. The bipolar world, with processes of evolution of international systems hampered or slowed and made latent, coerced theoreticians to seek for simple analogies and understandable metaphors. They were just at hand, accessible in cybernetics.

Second, theory of international relations, especially realism and neorealism , required a concept similar to equilibrium existing in neoclassical economics. On one hand such a concept had to be intuitively comprehensible and on the other, it had to be useful in mathematical modelling. In economics *equilibrium* proved to be very useful both in modelling and in producing heuristically valuable analogies and metaphors, e.g., market equilibrium. In international relations theory, stability in its narrowed sense was useful in arms race modelling. In other cases it only hampered methodological development by drawing too much attention to 'linearized', deterministic cybernetic analogy.

Turbulence and Chaos

Turbulence in world politics is understood as high complexity, dynamism and interconnectedness [42a]. It is then treated not as a metaphor but as an analytical concept allowing the study of three levels at which the world politics is influenced: individuals, collectivities and at the mix of the two levels. Such an approach emphasizes uncertainty of whole processes although with ordering processes based on the concepts of 'non-equilibrium thermodynamics'.

Deepened analysis of the ideas of Rosenau exceeds the scope of the paper. Nonetheless, in spite of efforts to define turbulence as an analytical concept [42a], a closer look at its explication reveals its dominant descriptive metaphorical meaning.

The term is associated with the analogy of the world system as an extremely intricate, fluctuating, 'turbulent dynamic' system. However, the term turbulence, associated with irregularity, tumult, insubordinance and riots, is a dormant metaphor both in physics and in everyday language. In Latin *turbulentus* means stormy, violently disturbed, irregular and *turba* could mean crowd, disorderly mass of people [37b]. Thus when used to describe social systems, turbulence returns to its initial meaning depicting disordered behaviour of people. In that case the metaphor is made alive although in a different sense—humanity can be viewed as a large more or less organized crowd.

The importance is that Rosenau's insightful proposal of a new approach to world system dynamics is not achieved due to analytical capacity of the concept of turbulence. The definition of complexity used in that case, deeply rooted in classic cybernetics and automatic control theory, e.g., 'organized' and 'disorganized' complexity [51] is supported by intuitively comprehensible patterns of dynamics proposed by 'non-equilibrium' thermodynamics and chaos theory. That idea helps to create an analogy between the world system and a highly complex, yet *classic* cybernetic system. The whole concept is finally made more convincing by the *metaphor* of turbulence.

Complexity of a social (international) system, is always the outcome of observation and self-reference [29]. Moreover, it is very difficult to prove close relevance between social systems and thermodynamic systems permitting to claim that anything more but sophisticated analogies and metaphors can be used. They can be useful as 'analytical' tools yet first their whole meaning is to be described: source and target domains, their limitations and validity.

A similar discussion can be illustrated with other uses of the language of chaos theory, non-equilibrium thermodynamics in grasping the dynamics of social systems. Basically they can be divided into two groups: (a) mathematical modelling of international systems [45] [52]; (b) new, paradigmatic views of dynamics of various social systems, including the world system [22] [25] [10] [11] [35] [49]. The concepts belonging to the second group depict social systems (the world system) as 'living ones', self-organizing, learning, thanks to the fluctuation far from equilibrium by means of 'autocatalytic' processes [35]. In the period of changes it is becoming very unstable and in some works efforts are made to show *explicitly* such transition processes as bifurcations, e.g., [25] [11].

Metaphorical meaning of chaos allows to address not the analytical meaning of those concepts but just the way they are created. According to The Oxford English Dictionary, chaos in its literal sense means '... any vast gulf, chasm, the nether abyss, empty space, the first state of the Universe' while in figurative sense it means: 'A state resembling that of primitive chaos, utter confusion and disorder ... where the parts are undistinguishable' [48]. Thus in everyday language and in mathematics, chaos is a dead metaphor, although with differing image-producing fields.

Concepts of chaos, non-equilibrium and partly turbulence draw our attention not to the sense of metaphors but to the meaning of analogies between social systems and various non-equilibrium source systems. Instead of refining 'analytical' and 'objective' validity of analogies by extended mathematization as in economics, cf. [44], it is possible to analyze which aspects of those analogies and associated metaphors better correspond with our perception of change in social systems. Preliminary evaluation shows that those analogies and metaphors allow to grasp the main features of contemporary social and international systems typical for a 'communication age' world, e.g., differentiation, self-organization, etc. There are, however, two more important aspects of those analogies and metaphors that seem equally significant.

First, 'technocratic' analogies and metaphors associated with classic cybernetics - stability and equilibrium, concealed emotional bearings of dynamics of international systems, or at least, exposed its appeasing symptoms. Additionally they treated stability as a result of purposive action, for instance through deterrence. Analogies and metaphors connected with modern systems thinking—turbulence, chaos, non-equilibrium relate to other emotions: anxiety, fear, uncertainty, and frustrate the prospects for purposive, concerted human activities. Reminiscence of the specific normative features of those metaphors and analogies allows to contend that in applications of modern systems concepts in social sciences, anxiety and fear are sanctioned as a part of 'objective' and 'rationalist' elements of social reality.

Second, the same analogies and metaphors, due to the same and other features, e.g., emphasis of differentiation and self-organization seem very relevant and supportive, also in a 'rationalist' manner, for liberal-democratic systems and market economy, cf. [54] [20]. In turn, analogies and metaphors based on classic cybernetics could be regarded both as denying central planification, e.g., the 'Law of Requisite Variety' [2], or as an appropriate foundation for super-planification [1].

CONCLUSIONS

Two basic conclusions can be drawn from the above discussion although more questions still remain unanswered.

First, in most of the applications of concepts taken from systems thinking, too little attention has been paid to their interpretations as descriptive analogies and metaphors. It brought about double consequences. In theoretical research many futile efforts have been made to make them more 'scientific', 'objective' and 'analytical'. Practice, in turn, has been enriched with 'objective' terms with various hidden normative content.

Some of these problems can be avoided if the following phenomenon is taken into account. It can be ironically called 'The Law of Metaphorical Infinity of Social (International) Systems':

> *'Any theory and/or model elaborated in physics, chemistry, biology, automatic control theory, in order to study collective phenomena ('systems thinking'), can be applied as a source of analogies and metaphors in various attempts aiming at description, explanation (sometimes even prediction and prescription), of processes taking place in social systems, beginning from small groups, and ending with the world system.'*

Second, along with the differentiation, pluralization, multipolarity, etc., observed in the contemporary world together with conviction about their inevitability and efficiency - 'The End of History' [15]; more interest is paid to concepts drawn from systems thinking, which facilitate depiction and explanation of that reality. However, stress must be put not only on 'objective' soundness of those concepts. Instead, they should be seen as analogies and metaphors allowing to reflect normative, emotionally-loaded understanding of the world. Thus the studies which intend to deal with the old and new systems-thinking related terms: autopoiesis, chaos, complexity, equilibrium, fluctuation, homeostasis, non-equilibrium, self-organization, stability, synergy, turbulence, ultrastability, should first concentrate on their semantic aspects, and later, try to use them in depicting social (international) dynamics.

Another argument supporting the last recommendation can be drawn from recent applications of ideas already well-established in international studies like 'anarchy' and 'security' [8] [53]. Emotionally laden concepts are made multi-dimensional and 'analytic'. Inevitably they will be amalgamated in international relations studies with old and new systems analogies and metaphors. Therefore, for future research, the following admonition is proposed.

Several of the above mentioned terms can be combined into more or less heuristically valuable pair-wise clusters like 'secure stability', 'stable security', 'secure chaos', 'chaotic security', 'stable chaos', 'stable anarchy', 'secure turbulence', 'turbulent security', etc. In some instances they seem productive for the future reflections on peace, security and change in the post-Cold War world. Nevertheless, before making any 'technocratic' analyses aiming at far-reaching generalizations, it is necessary to see them just as a kind of specific analogies and metaphors enriching the 'poetry' of social sciences.

The main question challenging further research on the use of systems analogies and metaphors in international relations studies, or in social sciences in general, can be summarized as follows.

Social systems can be treated as self-referential and autopoietic systems of meaningful communication. Thus they are created by language dominated by various metaphors and analogies [29]. Therefore the systems analogies and metaphors used in the language of social sciences (international relations studies) are bearing a specific appeal. First, their normative value is enhanced, as it was discussed before, and second, viewed as an instrument of self-reference, they contribute to the development of the system. It will then be necessary to study what the role is of all systems metaphors and analogies in the self-reference and autopoiesis of those systems. We have already proved that the first cybernetics-based analogies and metaphors strengthened the will and perhaps the possibility to control social systems, e.g., stability: repeating that the system is stable contributes to its stability.

The question is getting more intricate when we take into consideration analogies and metaphors derived from chaos theory and autopoiesis. It will be very productive to study to which extent analogies and metaphors of self-organization, self-reference, chaos and autopoiesis viewed as an element of self-reference, self-organization and autopoiesis contributes to the dynamics of social systems. In other words, we have to study whether those metaphors and analogies contribute to self-organization and autopoiesis, or maybe, their impact is quite opposite.

The question is very general, and of course very speculative. The first answer that we can give seems, however, not so speculative in the light of the above discussion. In a period of a decreasing role of mathematical modelling and in the time of growing meaning of discourse, relativism, or in other words, in the post-modernist world, the approach based on studies of the impact of systems analogies and metaphors is one of the most promising methods of investigation of dynamics of international systems. We may expect that the proposed preliminary ideas will provide us with more productive descriptions and explanations, and, perhaps even with insights of some predictive ability. It is of very great importance in the time after the collapse of the simple world of the Cold-War. At present we observe an unprecedented triumph of liberal democracy: 'The End of History'. At the same time, for societies, politicians and scholars this 'ultimate state of the world affairs' appears to be chaotic, turbulent and, perhaps, autopoietic.

Czesław Mesjasz
Academy of Economics &
Jagiellonian University
Cracow, Poland
e-mail: mesjaszc@ae.krakow.pl

REFERENCES

[1] **Afanasyev, V.G.**, *The Scientific Management of Society*, Progress, Moscow, 1977.

[2a] **Ashby, W.R.**, *An Introduction to Cybernetics*, Wiley, New York, 1963, p. 73.

[2b] *ibid.*, p. 81.

[3] **Axelrod, R.**, *The Evolution of Cooperation*, Basic Books, New York, 1984.

[4] **Beer, S.**, *Diagnosing the System for Organizations*, Wiley, Chichester, 1985.

[5] **Bellman, R.**, *Stability Theory of Differential Equations*, McGraw Hill, New York, 1953.

[6] **Bertalanffy, L. von**, *General Systems Theory*, Braziller, New York, 1968.

[7] **Brecher, M. & Wilkenfeld, J.**, *Crisis, Conflict, and Instability*, Pergamon Press, Oxford, 1989, pp. 29-30.

[8] **Buzan, B.**, *People, States and Fear. An Agenda for International Security Studies in the Post-Cold War Era*, Harvester Wheatshaf, London, 1991.

[9] **Cusack, T.R. & Stoll, R.J.**, *Exploring Realpolitik: Probing International Relations Theory with Computer Simulation*, Lynne Rienner, Boulder, CO, 1990.

[10] **De Greene, K.B.**, "Supplementary Systems Paradigms for Different Stages of Societal Evolution with Special Reference to War and Peace", *Systems Research*, **7**, 1990, p. 77.

[11] **De Greene, K.B.**, "The Turbulent-Field Environment of Sociotechnical Systems: Beyond Metaphor", *Behavioral Science*, **35**, 49, 1990.

[12] **Der Derian, J. & Shapiro, M.J.** (eds.), *International/Intertextual Relations: Boundaries of Knowledge and Practice in World Politics*, Lexington Books, Lexington, MA, 1989.

[13] **Deutsch, K.W. & Singer, J.D.**, "Multipolar Power System and International Stability", *World Politics*, **16**, 1964, p. 390.

[14] **Eatwell, J.; Milgate, M. & Newman, P.**, *The New Palgrave. A Dictionary of Economics*, Macmillan, London, 1987.

[15] **Fukuyama, F.**, *The End of History and the Last Man*, The Free Press, New York, 1992.

[16] **Gilpin, R.**, *War and Change in World Politics*, Cambridge University Press, Cambridge, 1981.

[17] **Gleick, J.**, *Chaos. Making a New Science*, Viking, New York, 1987.

[18] **Goldmann, K.**, *Change and Stability in Foreign Policy. The Problems and Possibilities of Detente*, Princeton University Press, Princeton, 1988.

[19] **Guetzkow, H. & Valadez, J.J.** (eds.), *Simulated International Processes: Theories and Research in Global Modelling*, SAGE, Beverly Hills, CA, 1981.

[20] **Hayek, F.A., von**, *The Fatal Conceit: The Errors of Socialism.* Collected Works of F.A. von Hayek. vol. I, Routledge, London, 1988.

[21] **Herman, E.S.**, *Beyond Hypocrisy. Decoding the News in an Age of Propaganda—Including the Doublespeak Dictionary*, South End Press, Boston, MA, 1992.

[22] **Jantsch, E.**, *The Self-Organizing Universe. Scientific and Human Implications of the Emerging Paradigm of Evolution*, Pergamon Press, Oxford, 1980.

[23] **Kaplan, M.**, *System and Process in International Relations*, Wiley, New York, 1957.

[24] **Kaplan, M.**, "Balance of Power. Bipolarity and Other Models of International Systems", *American Political Science Review*, **51**, 1957, p. 297.

[25] **Kast, F. & Rosenzweig, J.E.**, *Organization and Management. A Systems Approach*, McGraw Hill, New York, 1974.

[26] **Kegley, C.W. & Raymond, G.A.**, "Must We Fear a Post-Cold War Multipolar System?", *Journal of Conflict Resolution*, **36**, 1992, p. 573.

[25] **Laszlo, E.**, *Evolution: The Grand Synthesis*, Shambhala New Science Library, Boston, MA, 1987.

[26] **Liska, G.**, *International Equilibrium: A Theoretical Essay on the Politics and Organization of Security*, Harvard University Press, Cambridge, MA, 1957.

[27] **Lorenz, E.N.**, "Deterministic Non-Periodic Flow", *Journal of Atmospheric Science*, **20**, 1963, p. 130.

[28] **Loye, D. & Eisler, R.**, "Chaos and Transformation: Implications of Non-Equilibrium Theory for Social Science and Society", *Behavioral Science*, **32**, 1987, p. 53.

[29] **Luhmann, N.**, *Essays on Self-Reference*, Columbia University Press, New York, 1990, p. 81.

[30] **Mandelbrot, B.B.**, *The Fractal Geometry of Nature*, Freeman, Oxford, 1982.

[31] **Meadows, D.H. et al.**, *The Limits to Growth. A Report for the Club of Rome's Project on the Predicament of Mankind*, Universe, New York, 1972.

[32] **Meadows, D.H. et al.** (eds.), *Groping in the Dark: The First Decade of Global Modelling*, Wiley, New York, 1982.

[33] **Mesarovic, M.D. & Pestel, E.**, *Mankind at the Turning Point*, E.P. Dutton, New York, 1974.

[34] **Mesjasz, C.**, "Applications of Systems Modelling in Peace Research", *Journal of Peace Research*, **25**, 1988, p. 291.

[35] **Modelski, G.**, "Is World Politics Evolutionary Learning?", *International Organization*, **44**, 1990, p. 4.

[36] **Niou, E.M.S. et al.**, *The Balance of Power. Stability in International Systems*, Cambridge University Press, Cambridge, 1989.

[37a] *Oxford Latin Dictionary*, Clarendon Press, Oxford, 1968, p. 1812.

[37b] *ibid.*, pp. 1990/1993.

[38] **Prigogine, I.**, *From Being to Becoming*, Freeman, San Francisco, 1980.

[39] **Prigogine, I. & Stengers, I.**, *Order out of Chaos*, Bantam, New York, 1984.
[40] **Richardson, L.F.**, *Arms and Insecurity*, Boxwood Press, Pittsburgh, PA, 1960.
[41] **Rosecrance, R.N.**, *Action and Reaction in World Politics*, Little Brown, Boston, MA, 1963.
[42a] **Rosenau, J.N.**, *Turbulence in World Politics. A Theory of Change and Continuity*, Princeton University Press, Princeton, 1990, p. 65.
[42b] *ibid.*, p. 51.
[43] **Rosenau, P.**, "Once Again Into the Fray: International Relations Confronts the Humanities", *Millenium*, **19**, 1990, p. 83.
[44] **Rosser, J.B.**, *From Catastrophe to Chaos: A General Theory of Economic Discontinuities*, Kluwer, Boston, MA, 1991.
[45] **Saperstein, A.M.**, "The "Long Peace"—Result of A Bipolar Competitive World", *Journal of Conflict Resolution*, **35**, 1991, p. 68.
[46] **Scheffran, J.**, *Verification and Stability. The Strategic Impact of Uncertainty and Perceptions*, in **Brauch, H.G.** (ed.), *Verification and Arms Control: Implications for European Security*, Selected Papers, Part II, AFES-PRESS, Mosbach, 1990, p. 190.
[47] **Schelling, T.C.**, *Arms and Influence*, Yale University Press, New Haven, CT, 1966.
[48] **Simpson, J.A. & Weiner, E.S.C.**, *The Oxford English Dictionary*. 2nd ed., Clarendon Press, Oxford, 1989, pp. 22-23.
[49] **Wallerstein, I.**, *Unthinking Social Science. The Limits of Nineteenth Century Paradigms*, Polity Press, Cambridge, 1991.
[50] **Waltz, K.N.**, "The Stability of a Bipolar World", *Daedalus*, **93**, 1964, p. 881.
[51] **Weaver, W.**, "Science and Complexity", *American Scientist*, **36**, 1948, p. 536.
[52] **Wolfson, M. et al.**, "Chaotic Behaviour in International Conflict", *Journal of Conflict Resolution*, **36**, 1992, p. 119.
[53] **Wæver, Ole et al.**, *Identity, Migration and the New Security Agenda in Europe*, Pinter Publishers, London, 1993.
[54] **Zeleny, M.**, "Spontaneous Social Orders", *International Journal of General Systems*, **11**, 1985, p. 117.

MARÍLIA BERNARDES MARQUES

CONFRONTING HEALTH TRANSITION COMPLEXITY

INTRODUCTION

Dramatic changes in society, in technology and in the environment are conducting humankind to a new era of confrontation with death and disability and are favoring the emergence and dissemination of new pathogenic agents, broadening the spectrum of infectious diseases [1] [2].

Emerging and resurgent infectious diseases constitute new threats to public health and are contributing in a significant way to accelerate the spiraling costs of health care in diverse developed and developing countries. Continuously, episodes of emerging and resurgent infections are reported. Diseases like HIV/AIDS, the dissemination of dengue virus and the frequent appearance of previously unknown diseases like hemorrhagic fever and the rising of tuberculosis and cholera portray this trend.

Currently, results of biomedical research are originating new promising biotechnologies potentially useful to recognize and control infectious and parasitic diseases.

In a previous paper, Possas and Marques [3] discussed to which extent the emergence and global dissemination of new and resurgent infectious diseases are contributing to increase the complexity of health profiles and how this trend can affect the predictability of dominant health transition approaches . In that paper we suggested that a comprehensive transdisciplinary approach is necessary, incorporating the theories of complexity in ecology, in evolutionary theory, in social sciences and in information sciences. We also suggested that this new approach will provide the elements for the formulation of a theoretical framework for health transition.

Taking into account that emerging and resurgent infectious diseases and new health technologies are simultaneous and contemporary phenomena, the purpose of this paper is to analyse the possible existing relations between two different notions operating in the same time dimension—the future: prediction of disease and the technology prospective.

© 1999 *Kluwer Academic Publishers. Printed in The Netherlands.*

THE PROBLEM: NEW INFECTIOUS DISEASES

Scientific literature defines emerging and resurgent diseases as a known or unknown infectious disease, whose incidence in humans has increased in a recent period of time or that threatens to increase in the near future [4]. Frequently, the expression "new infectious diseases" is also employed to name this set of diseases.

In Brazil, besides the recrudescence of malaria in mining areas, of dengue, of HIV/AIDS, of cholera, other viruses are also described, among them: Oropouche Fever, Rocio encephalite. Recently, a new arenavirus was described, named Sabiá, in a fatal case of hemorrhagic fever, which received the initial diagnosis of Yellow Fever [5].

The emergence of Lyme disease, unknown in the US in the past two decades, was partially a result of reforestation and of an increasing amount of deers in the American Northeast. Currently it is the commonest tick-borne disease in that country, causing high morbidity rates in suburban and rural areas [6].

In the US, the CDC (Centers for Disease Control and Prevention) in a recent report related, among others, the following examples of new infectious diseases: *Escherichia coli* O157:H7, human mycoses, multidrug-resistant pneumococcal disease, hantavirus infections.

In the past 20 years, fungal infections have increased dramatically, and they are now important causes of morbidity and mortality causes of hospitalized patients. Partially, this increase was a paradoxical consequence of medical advances (prolonged use of broad-spectrum antibacterial agents, organ transplantation, cancer chemotherapy, invasive procedures like catheters and prothetic devices)[7].

New infectious diseases are stimulating discussions of old political subjects, such as possible links among genetic engineering, new pathogenic agents and biological weapons [8], microbial traffic [9], disease interchanges between the New and the Old World [10]. They are also revitalizing the hope and the claims that international solidarity can guide and inspire technology transfer and international scientific and technological cooperation [11] [12] [13].

THE NEW, THE UNEXPECTED, THE COMPLEX AND THE UNPREDICTABLE

In the US, the members of the Harvard Working Group on New and Resurgent Diseases [2], suggest that any effective analyses of emerging diseases should recognize the study of complexity as the main scientific challenge of the present time.

This assessment is supported by the idea that categorizing a new specific disease should necessarily encompass the analysis of the net of causes, with all its complexity. The final goal of this approach, which is being formulated, is to anticipate the appearance of new infectious diseases and to eliminate those conditions which facilitate the dissemination. These are challenges confronting a high degree of uncertainty.

The Harvard Group proposal comprises a type of strategic research, including ecological and evolutionary surveillance. This strategy should allow, at the same time: the prediction of infectious diseases in the short-term, through the identification of those conditions favoring outbreaks of diseases already known and through the anticipation of ecological changes associated with human activities; in the long-term, through modeling activities based in the evolutionary ecology.

This approach assumes the emergence of new infectious diseases as an evolutionary process, in which diverse natural and social processes interact [14]. Undoubtedly, emerging infectious diseases are inspiring a new and modern transdisciplinary approach.

Complexity Theory is related to Chaos Theory, which in the 1970s became a transdisciplinary viewpoint incorporating the system's multidimensional and global features. Chaos Theory became the study of systems that can be modified, in infinite and frequently unpredictable ways, by minimal inputs or changes. We consider that it is the *unpredictability* that, partially, allows us to recognize complexity.

One of the central assumptions of Complexity Theory is that complex adaptive systems can be quantified, using mathematical and physical models and computation sciences. Systemic Modeling does not include the concept of causal explanation or evidence. While Analytical Modeling is valuable to the comprehension of predictable phenomena, according to Tarride, Systemic Modeling is projective and has been developed precisely to allow to go from the predictability pressured by calculation to unpredictability [15].

Tarride also argues that Systemic Modeling is based on a constructivist epistemological cement, very different from the post-positivist foundation of Analytical Modeling. Although irreducible to its components, complexity can be modeled, according to Systemic Modeling reasoning. Complex thinking is not dedicated to point the elemental or, on the contrary, to do a global simplification, where everything can be reduced to a simple unit.

Tarride also argues that is not the case of substituting a classical way of doing science by a new science. Complexity Theory identifies the need for approaching the contemporary problems in a different way, but complementary to the traditional one. Consequently, the approach of problems affecting ecosystems and humans, as persons and as social groups, is a major challenge to complexity thinking and to Systemic Modeling.

TIME, UNCERTAINTY AND RISK

Time is a variable difficult to deal with and the future is essentially uncertain, unpredictable. With Bifani [16] we also consider 'that it is very important to distinguish the social time rationality, which is of short-term, from the time of scientific and technological development, which is of long-term, and the time of the evolutive dynamic of natural systems, whose rationality is centenarian.

Time is a crucial element to the comprehension of the dynamics of human population development. Short-term rationality is predominant when the possibility of a social intervention is being considered. The results of the intervention, however, will be seen immediately and in the long-term.

We should take another pre-eminent notion in the comprehension of diverse dynamic process in health: the risk. A large amount of specialized literature is dedicated to the study of risk in health and medicine. Most of it is dedicated to the practical applications of risk, including, among other aspects, the assessment of medical products and procedures, of technological and natural disasters, and of health security plans [17].

Different authors agree that the methods and techniques contemplated in the formal analysis of risk in health, have a dominant logical-positivist approach, in which the final goal of the prediction is to decrease the number of occurrences, through the reduction of the uncertainty edges.

On the contrary, the time rationality in the dynamic of natural evolutive systems is essentially of very long-term. For each animal or vegetal species, evolution is based on the transmission of changed genetic information, through spontaneous mutations. In the evolutionary process, the intervals of time are long, centenary, secular.

Human prosperity has in the development of the brain, at the biological level, its main past evolutionary feature. In human species, cultural evolution makes possible the transmission of acquired experience between successive generations, through the use of spoken or recorded words.

In the past decades, computational techniques have enormously increased the human capacity to accumulate and communicate informations. These techniques are generating a revolutionary or paradigmatic leap in humankind's evolutionary process.

The result is that the human evolutionary process (of communication-interchange of informations) consumes much less time, in comparison with other biological species time-consumption. Hence, the development of human species is, predominantly, due to its cultural evolution and not an exclusive consequence of genetic evolution.

Frequently, the current ecological or environmental problems are considered basically as outcomes of the biological success of human species and of the dominant role of human action over diverse ecosystems and of its capacity to dominate the mechanisms of biological control. The accumulation of scientific information in the life sciences, at the same time that enabled humankind to manipulate genetic information, also contributed to the decrease of global biodiversity , favoring human intervention in natural ecosystems. Thus, the consequences of the successful human evolution were the qualitative and quantitative changes observed in two components—uncertainty and risk—in daily life.

Tiezzi's [18] reflections on this subject are relevant: human species has multiple relations with other species and its introduction in a natural system does not increase the complexity of the system. On the contrary, its introduction can result in the simplification of the natural system. To this author, environment damage, disordered increase of some species, and total destruction of some ecosystems, are caused by this simplifying human action. Considering that the outcomes of this simplifying human action are deforestation, desertification, soil erosion, degradation of irrigated lands, industrial and agricultural pollution, global warming, destruction of the ozone layer and species extinction, we argue: are emerging and resurgent infectious diseases outcomes of this simplifying human action on natural systems?

According to Tiezzi, the human being invented mutations for his adaptation to the environment and perverted his own biological time through the use of scientific and technological resources that he developed as a self-gift.

To support this conjecture, we argue that it is enough to refer to the techniques of artificial fertilization, the liberalization of microorganisms genetically modified in the environment, the engineered varieties of plants.

For Tiezzi, ecology is the science of complexity and the concept of time has an important role in the definition of complexity. For this author, the best way for the survival of living systems on earth may not be to favor future human being complexity but, on the contrary, to preserve the ecosystems complexity.

Finally, we should consider the dynamic of scientific and technological development, whose rationality is essentially long-term.

The social and environmental dimensions in the impact assessment of science and technology frequently conduct us to confront a set of uncertainties and risks, of great magnitude. We also know that although the specificity of each technological innovation life cycle, the time interval between an invention and the process of technological diffusion is, in general, large, achieving in some cases some decades. Time, uncertainty and risk thus constitute some of the main variables in the analysis of the scientific and technological development process.

The element time is crucial when we are taking into account the possibility of intervention through the formulation of policies, planning, managing and assessing science and technology. All these are activities focusing the future.

On the other hand, the direct and the indirect outcomes of technological innovation can be immediate or be noticed only in the long term. The outcomes of technological application (diffusion) in industry or in the agricultural sector, are not limited to increasing productivity. They can also encompass diverse qualitative and quantitative changes in society (social change) and nature (ecological change).

According to Bifani [16], the application (diffusion) of science and technology in the social system and in nature results in a large spectrum of direct and indirect effects in different functional and physical spaces. These effects also become materialized in different horizons of time. Social system and nature are not passive, reacting to this diffusion. The consequence is that the original intervention of science and technology starts a complex dynamic of effects and reactions. These effects are increasingly large in intensity and time and spread in the space, with synergic phenomena amplifying the final impacts.

As a consequence, to predict the probability of occurrence (risk) of these effects and reactions, and of their magnitude as well, is a very difficult task, confronting increasingly uncertainties. Unexpected effects and reactions which are not necessarily negative, can be obscure and permeate the entire social, economic, political and cultural structures, and nature as well. They usually scape or are not considered in the decision-making processes, that is, in the formulation of policies, planning, management and assessment of science and technology.

To confront this undesired situation and to avoid negligence with hidden phenomena, it is important to try to make explicit the risks of occurrence of the unexpected event. Thus, risks start to be considered as inherent to any decision-making process.

The definition of research priorities and the technological choice should represent, therefore, *anticipation* activities, capable to take into account the risks of occurrence of unexpected outcomes and to announce the best investment opportunities as well.

MARKING THE DIFFERENCES BETWEEN PREDICTION AND THE PROSPECTIVE

Images of the future are scenarios, forecastings. How to study the future and to anticipate what the future signifies for the human being?

The Canadian futurist Norman Henchey, quoted by Hancock & Bezold [19], describes four manners of thinking the future: The possible future; The plausible future; The probable future; The preferable future. Thus, while the probable future scenario fosters the option of preparing to confront it, the preferable future scenario originates the expectation that the future is a value to be created.

According to Godet [20], the future is the "raison d'être" of the technological prospective. This author considers that the future should be studied in a global approach, that can illuminate the present. This constitutes the central idea that differentiate a prospective from a mere prognosis. The prospective is a critical viewpoint of the future, that should direct actions in the present and is not barely an exercise of futurology.

The prospective is a manner of thinking to act against fatalism, which gives the means to understand and to explain the crisis, in a world characterized by increasing uncertainty and risk and by changing trends, not only in the long term but also in the short term.

The technological prospective is more formal, in comparison with the activity of prediction of diseases. It is processed in a more specified framework of time and contingencies and is concerned with diverse techniques of anticipation. Its purpose is not to produce prognosis, but conjecture. It is an attempt to outline and analyse the plausibility of a certain number of different scenarios and to construct conjectures about them, with a practical objective: to support decision-making, the choice, the definition of priorities.

Hence, the activity of the technological prospective is associated, at a more general level, to policy formulation and to strategic management, in public and private enterprises.

Diverse formal procedures such as building scenarios (relevance trees), expert consensus methods (Delphi, specialists panels, brain storm sessions), structural analysis, morphological analysis, cross-impact analysis, matrix representation, techniques of prognosis, simulation, contextual mapping, decision-trees, among others, constitute techniques of the scientific and technological prospective.

Eventually, these techniques can be useful to clarify and to explain the risks and the opportunities profiled in the long term, inasmuch as it is considered that the future cannot be anticipated with absolute certainty.

Prospective outlines and analyses of a certain number of scenarios or *futuribles*, assessing the possible future evolution of a situation, phenomenon or problem, preserving a certain distance, can be done.

The method of scenarios consist of two phases: the construction of the data-base and from this data-base the construction of the scenarios. Projections will result from the constructed scenarios. After this prospective activity follows the strategic planning, which includes the definition of strategies and the choice of strategic options.

To sum up, in both tasks—prediction of diseases and technological prospective—it is the case of looking in the direction of the future. However, although both comprise interdisciplinary tasks, if prediction conduct is prognostic, the prospective produces conjectures on the future. While the prospective is related to anticipation, prediction needs a degree of certainty hardly found in the reality.

Arocena [21] considers that in the populational prediction studies, the formulation of causal hypothesis and the selection of diverse factors should constitute the decisive step while formal mathematical representation and calculation procedures have a subordinate importance.

Rattner [22] points some of the future studies drawbacks: most prediction models presuppose a perfect knowledge of reality. This is conceived as a system whose elements are necessarily in harmony and whose dependent and interdependent variables and the respectives patterns of relations, would be controlled by the owners of the power. As a matter of fact, the elements composing the system are, in general, in tension and conflict and it is doubtful that it is possible to achieve a plain knowledge of all the variables that determine its functioning and the direction in which the system develop.

Arocena also argues that the confrontation of predictions done in the past with current populational reality, shows discrepancies which are linked to the presence of the uncertain, like migration, or certain scientific and technological health advances, that can surprise, with their impacts surpassing expectations or, on the contrary, reject the hypothesis.

This critical perspective conducted to the identification of particularities and individualities and, thanks to it, nowadays it is perfectly understood that it is impossible to establish conjectures on the uncertain, on the new, based in a linear demographic prognostic. Arocena gives some examples: Sweden and other North European countries, shows an unanticipated fecundity growth. This is evidence that higher developed societies are returning to traditional family patterns, despite all the dominant linear predictions of the demographic transition. Regarding mortality, there are hypotheses of life expectation of 125 years simultaneous with catastrophic assumptions linked to the environmental and climate shadowy perspectives and new infectious diseases as well. In fact, the emergence of a new infectious disease among humans, representing an unexpected adverse event, frequently is surrounded in the specialized literature and in the press, by fatalism and anticipations of epidemiological disasters and other tragedies. Together with new diseases in animals and plants, the entire set of new infectious diseases would be introducing global risks to world economy and food availability, and to the equilibrium of ecology.

It is not possible, however, to predict with absolute certitude, future demographic and epidemiological trends. It is impossible to have 100% accuracy in predicting the future course of fecundity, mortality, migration and events like the emergence of new infectious diseases.
It is this uncertainty that makes viable to outline alternative plausible scenarios and develop a more realistic prospective analysis of the future, based in the knowledge of the present. These prospective analyses, allowing to anticipate different possibilities, can contribute to avoid them or, on the contrary, to make them a reality.

Hence, because populational developments are inherently uncertain—and migration phenomena are especially unpredictable events—the alternative prospective scenarios can be a valuable resource. The use of prospective scenarios and other alternative methods and techniques of the prospective are a resource to deal with uncertainty. In front of the uncertain, we are proposing the introduction of prospective analysis in epidemiology.

It is accepted in general that the concept of epidemiological transition should orient the planning of the type and scale of health necessities in the future, in different countries. Nevertheless, we argue that it will be necessary to surpass in the health field, the ordinary futuristic and deterministic predictions, which also are, frequently, catastrophic and based in static epidemiological profiles and in analysis of past and present trends of mortality and fertility.

We also argue that the introduction of anticipations or projections with a reasonable sensitivity, in epidemiology, in an activity of construction of alternative epidemiological scenarios, based in diverse political, social, market, and cultural conjectures, could direct health care planning and scientific and technological activity in health, as well.

We consider that Systemic modeling can offer the necessary physical support to the construction of insightful alternative scenarios or futuribles in the field of uncertainties surrounding new and resurgent diseases.

We argue that other resources valuable in the study of new infectious diseases are the Geographic Information Systems (GIS), encompassing geo-processing techniques, satellite and remote sensing technology. The introduction of these physical supports in the health field, although in a very initial stage, has already proved the relevance of the GIS to the renewing of medical geography. Especially hopeful are the recent efforts to define more precisely the topographic and meteorological conditions associated with the abundance of some parasites [23] [24].

Political ecology of disease represents an enlargement of causal thinking. Its concept of cause is multifaceted, including behaviour, environment, culture, society and pathogenic agents. This open-minded way of examining diseases also places diseases prediction as one of the main goals of the health political ecology. Thus we propose that this field also incorporate the prospective. We also propose the association between science and technology prospective assessment and health strategic planning. The last should encompass the time interval extending from the invention or discovery, for example, of a new anti-microbian agent, to the research and development (R&D) of a new drug or vaccine, to the industrial scale-up, commercialization, regulation and incorporation to health care services.

The incorporation of technologies in health services requires the future fulfillment of certain conditions or requires the correction of distortions (medical spiraling costs) and the surpass of present health crisis situation, related to the Welfare State crisis.

The prospective analysis should incorporate the central question of democratization of scientific knowledge. By taking into account individual and public interests, in the long-term, the prospective analysis should have the capacity of impulsing open and pluralistic dialogues about major social options linked to the future.

Finally, we consider that the prospective analysis, thanks to its capacity of enlarging public comprehension on the ethical and safety risks of science and technology diffusion, should enrich the relations among science, technology, nature and humankind.

Marília Bernardes Marques
Oswaldo Cruz Foundation
Rio de Janeiro, Brasil

REFERENCES

[1] **Institute of Medicine** (ed), *Emerging Infections: Microbial Threats to Health in the United States*, National Academy Press, Washington, DC, 1992.

[2] **Levins, R.; Tamara, A.; Brinkmann, U; Eckardt, I; Epstein, P.; Makhoul, N.; Possas, C.A.; Puccia, C.; Spielman, A. & Wilson, M.**, "The emergency of new diseases", *American Scientist*, **82**, 1994, pp. 52-60.

[3] **Possas, C.A. & Marques, M.B.**, "Health transitions and complex systems: a challenge to prediction?", In: **Wilson, M.E.; Levins, R. & Spielman, A.** (orgs.), *Disease in Evolution. Global Changes and Emergence of Infectious Diseases*, Annals of The New York Academy of Sciences, New York, vol. **740**, 1994, pp. 285-296.

[4] **CDC** (Centers for Disease Control and Prevention), *Addressing Emerging Infectious Disease Threats: a prevention Strategy for the United States*, Georgia: NCID/CDC, 1994.

[5] **Marques, M.B.**, "Emerging infectious diseases in the realm of complexity: implications for scientific and technological policies", *Cad. Saúde Públ.*, Rio de Janeiro, **11**, 3, 1995, pp. 361-388.

[6] **Barbour, A.G. & Fish, D.**, "The biological and social phenomenon of Lyme disease", *Science*, **260**, 1993, pp. 1610-1616.

[7] **Georgopapadakou, N.H. & Walsh, T.**, "Human mycoses: drugs and targets for emerging pathogens", *Science*, **264**, 1994, pp. 371-373.

[8] **Geissler, E.**, "Strengthening the biological weapons convention by the biesenthal vaccine initiative", In: **Tzotzos, G.T.** (org.), *Biotechnology R&D Trends. Science Policy for Development*, Annals of The New York Academy of Sciences, New York, 1994, pp. 43-52.

[9] **Morse, S.S.**, *Emerging viruses and their causes*, The Rockefeller University (commissioned review article, mimeo), New York, 1993.

[10] **Berlinguer, G.**, "The interchange of disease and health between the old and new worlds", *International Journal of Health Services*, **23**, 1993, pp. 703-715.

[11] **Jain, S.C.**, "Global Health: Emerging Frontier of International Health", *Asia Pacific Journal of Public Health*, **5**, 1991, pp. 112-114.

[12] **Doll, R.**, "Health and the environment in the 1990s", *American Journal of Public Health*, **82**, 1993, pp. 933-941.

[13] **FAS** (Federation of American Scientists), *Public Interest Report*, **46**, 6, nov/dec, 1993.

[14] **Eckardt, I.**, "Challenging complexity. Conceptual issues in an approach to new diseases", In: **Wilson, M.E.; Levins, R. & Spielman, A.** (eds), *Disease in Evolution. Global Changes and emergence of Infectious Diseases*, Annals of The New York Academy of Sciences, New York, vol. **740**, 1994, pp. 408-417.

[15] **Tarride, M.I.**, *Complejidad y Sistemas Complejos, The National School of Public Health*, Oswaldo Cruz Foundation, Rio de Janeiro, Brazil (Paper presented to the Doctorade Program), 1994.

[16] **Bifani, P.**, "Cambio tecnológico y politica cientifica y tecnologica", In: **Martinez, E.** (org.), *Estrategias, Planificación y Gestión de Ciencia y Tecnologia*, Cepal-ILPES/UNESCO/UNU/CYTEd-D, Editorial Nueva Sociedad, Caracas, 1993, pp. 99-123.

[17] **Hayes, M.V.**, "On the epistemology of risk: language, logic and social science", *Social Science and Medicine*, **35**, 1992, pp. 401-407.

[18] **Tiezzi, E.**, *Tempos Históricos, Tempos Biológicos. A Terra ou a Morte: os Problemas da Nova Ecologia*, São Paulo, Brasil, Nobel, 1988.

[19] **Hancock, T. & Bezold, C.**, "Possible futures, preferable futures", *Health Care Forum Journal*, 1994, pp. 23-29.

[20] **Godet, M.**, *Scenarios and Strategic Management*, Butterworth Scientific Ltda, London, 1987.

[21] **Arocena, R.**, "Acerca de la prospectiva (desde algunos países de América Latina)", In: **Martinez, E.** (org), *Estratégias, y Gestión de Ciencia y Tecnologia*, Cepal-ILPES/UNESCO/UNU/Cyted-D/Editorial Nueva Sociedad, Caracas, 1993, pp. 185-289.

[22] **Rattner, H.**, *Estudos do Futuro: Introdução á Antecipação Tecnológica e Social*, Fundação Getúlio Vargas, Rio de Janeiro, 1979.

[23] **Dobson, A. & Carper, R.**, "Biodiversity", *The Lancet*, **342**, 1993, pp. 1096-1099.

[24] **Epstein, P.R., Rogers, D.J. & Slooff, R.**, "Satellite imaging and vector-born disease", *The Lancet*, **341**, 1993, pp. 1404-1406.

PHILIP L. PETERSON

NON-NATURAL CULTURAL UNIVERSALS EXIST

SUMMARY AND CAVEAT

Significant linguistic universals exist and significant musical universals exist. Both are cultural universals. Significance via irreducibility to non-cultural biological phenomena rests in the centrality of knowledge-acquisition for each activity. Cultural universals are unreducible (non-natural) when part of the required knowledge remains non-innate. Surprises result from applying current neo-essentialism (Kripke)—e.g., words are like natural kinds and have essential properties only empirically discoverable (including meanings!); melodies also have essential empirical properties (Happy Birthday's dominant beginning and tonic end); and the locus of contingency is shifted to the decisions and actions of cultural participants.

The question has two parts: (i) do any cultural universals exist, contrary to the initial absurdity? and (ii) do these cultural universals play significant explanatory roles and resist reduction to 'nature'? The answer is yes, since significant linguistic universals exist and all linguistic universals are cultural universals. Similarly, significant musical universals exist and are additional cultural universals. Both kinds of universals issue from a recent variety of neo-rationalism wherein human knowledge acquisition presupposes innate knowledge whose content includes the universals. Although all cultural phenomena are biologically supported, irreducibility to non-cultural biological phenomena rests in the centrality of knowledge-acquisition for each activity. Cultural universals are unreducible when part of the required knowledge remains non-innate. Further, current neo-essentialism applied to cultural universals does not destroy significance, but the locus of contingency is shifted. It arises from the decisions and actions of cultural participants, rather than residing in cultural objects and phenomena in themselves.

All uses of the words 'culture' and 'cultural' herein concern 'culture' in what I have deemed the 'anthropological' sense rather than in what I have deemed the 'humanistic' sense [29]. I proposed that the first sense was the anthropological concept of culture—that expressed by a dictionary entry like 'the sum total of ways of living built up by a group of human beings and transmitted from one generation to another'—and that the second sense was the humanistic concept of culture—that expressed

© 1999 *Kluwer Academic Publishers. Printed in The Netherlands.*

by a dictionary entry like 'the quality in a person or society arising from an interest in and acquaintance with what is generally regarded as excellent in arts, letters, science, manners, morals, and scholarly pursuits' [1]. Now the caveat. Even though I am considering the anthropological concept of culture, that does not mean that I am taking an 'anthropological perspective' or that I am adopting the point of view of any anthropologist. Rather, I offer a philosophical analysis, albeit one concerned with what anthropologists do consider central to their discipline. 'To most anthropologists—at least in the United States—culture has been the discipline's core concept ... culture refers to those phenomena which account for patterns of behaving that cannot be fully explained by psychobiological concepts.' [18]. One of my aims is to defend the 'culture-nature' distinction, something many anthropologists and some philosophers may think impossible without bringing in the humanistic concept of culture—or even with it) [2].

EXISTENCE AND SIGNIFICANCE

To start with, it ought to seem absurd to say that cultural universals exist, absurd because it flies in the face of the evidence. For cultural properties are differences. Chinese culture is what it is because inter alia it is not contemporary Arabic culture. The differences define the cultures and what they have in common is not 'cultural' in the usual sense of the word; e.g., both Chinese and Arabs eat and reproduce, but those facts aren't facts of culture properly speaking—but of non-cultural, biological nature. For it's the different ways they do it which is cultural—what and with whom they eat, how social links are established that promote procreation. Granted cultures affect each other and sometimes overlap and so some genuinely cultural properties are shared (e.g., literateness, sometimes specific language). So, there might be, by sheer accident, some cultural feature or other that all cultures share; say, socially favoring the right-handed. But such 'cultural universals' could easily be insignificant for one or both of two reasons: (i) explanatory triviality (of no use in advancing general understanding of culture or nature); (ii) reducibility to 'natural' phenomena (i.e., biological, physical, physiological, or other non-cultural fact). Perhaps the main reason expected today for doubting the existence of cultural universals would be the multi-culturalism insisted on by so-called Postmodernists. (An ironist will find it easy to say that Postmodernism is simply a recent wrinkle in the time-honored tradition of French skepticism extending back to Montaigne, not to mention Protagoras) Any proposal for a cultural universal would be regarded as a manifestation of the dominant elite demonstrating their power to dictate the 'canon'. For this kind of skepticism, all reasons, principles, and theories are political—in short, Thrasymachus' view expanded beyond justice to truth, knowledge, beauty, and/or anything you like. If I were engaged in accounting for Postmodernist views on cultural universals (which I am not), then I would follow the lead of Kuspit [3] [21]. Expecting to find cultural universals would be like expecting to find linguistic universals, where the obvious observable fact about natural languages is that they are different from each other. French is not English, nor Marathi. Why? Because none of the words or sentences are the same. You might say that 'dossier' (in English) and 'drugstore' and 'weekend' (in French) refute that. But look at the pronunciation of those words. They are pronounced differently in French and English. Phonology is as central to a natural language expression as syntax and meaning. The French have made the English word 'weekend' their own word, as shown by its pronunciation (not to mention that they say *le* weekend).

However, contrary to this plausible attitude on cultures and languages, recent research has shown that looking for linguistic universals at the superficial level (concerning directly observable aspects of sound, form, and meaning) is not explanatorily fruitful. Looking somewhat below or behind the surface properties, however, does give rise to genuinely fruitful universals. Just consider syntax. Every natural language centrally contains nouns and verbs which play the most central roles as constituents of sentences of the language. Every language selects from a finite stock of humanly possible producible and audible sounds to use in constructing the sounds of expressions (e.g., the clicks of Xhosa are on this list, but bilabial whistlings are not). These facts about natural languages are facts about their similarities. They are, for the most part, not immediately observable on the surface, but they are crucial in that each particular language cannot be accurately described (much less explained) without mentioning and using them—together, of course with non-universal facts (e.g., that 'célibataire' is a noun in French).

Further than this, however, as well as beyond the much more fruitful non-superficial properties like 'having a grammar with a generative-transformational form' (or even 'being the one universal transformation Move-alpha'), the main role of universals in recent grammatical theory has been to meet the difficulties of explaining language acquisition. The central scientifically motivating phenomenon is the 'poverty of the stimulus'. How is it that the child learns so much with so little time and information? The hypothesis widely adopted is that the child already knows a good deal of what he seems to have to learn so that the overwhelming task is reduced for the language-learner to one that can be informationally managed. And what is innately known are linguistic universals. So, linguistic universals exist precisely because every human being not only tacitly knows them—but knows them innately! Recently, Chomsky ([16] [17]) and others have proposed a 'principles and parameters' revision of linguistic theory in which all the principles in question constitute universal grammar (UG) and when values of parameters in UG are set in certain ways the grammars for particular languages result. All the components of UG are linguistic universals.

The existence of any innate knowledge remains controversial in epistemology. However, I shall not herein present and defend an innatist knowledge theory, but shall simply assume that at least some knowledge is innate in adopting a brand of neo-rationalism heavily influenced by Chomsky and Kant. The main theoretical element that Chomskyan linguistic theory has added to this neo-rationalism is a rich concept of tacit, internalized knowledge. In sum, a natural-language user knows, in some appropriate non-figurative sense of great philosophical interest deserving more epistemological explication than it has received, the grammar of the language he possesses. In the most radical statement of this proposal, a speaker-hearer knowing his language is nothing other than his knowing the grammar of it in such a way that language and grammar are the same thing (cf. the 'internal' language, identical to the language's grammar [17]). Knowing the words of a language and what to do with them—that knowledge—has a content which can only be described by detailing what the linguistic facts are which must be known (unconsciously or tacitly) to use the language both in speaking and understanding—'use' it minimally, aside from all the complications of actual performance which go beyond bare knowledge of the language used. This tacitly possessed knowledge-of-the-language is not easily conceived of as a list of true propositions, or as any analogous components of a 'language of thought' (say), or indeed as anything that it is easy to conceive of as being directly noticeable "in consciousness". For example, try to be conscious of the rules for pluralization as you speak and make all the plurals correctly. Can you either notice the rules themselves in your consciousness as you pronounce your singular and plural nouns correctly, or notice that you are satisfying the rules just as you pronounce them? I think you don't and can't. It is only by reflection on your remembered conscious performance (the previous consciousness being what Sartre called 'pre-reflective') that you can be conscious of the rule or its having been applied. But that is not noticing the 'rule in action' (so to speak). I submit that you can't notice even such a simple rule as it operates because it is not a proper object of direct consciousness, even though it certainly is an aspect of mental performance and cognition. It's simply not an introspectible aspect and is only knowable after the fact, reflectively and, thus, by memory in part. (This bears on answering Kripkenstein, by the way.) It is not my topic to specify further how this content should be characterized epistemologically. I just assume it, while calling your attention to the fact that some of the content is innate and some of it is acquired. (That 'célibataire' is a French word for bachelors and that Spanish is a 'pro-drop' language is knowledge that has to be acquired, whereas many other linguistic features of French and Spanish are innate—viz., those features which are also properties of every other natural language.)

Since there are linguistic universals that aren't explanatorily trivial, there are cultural universals that aren't explanatorily trivial. For natural language is one of the central components of culture. If language isn't a cultural phenomenon, what is? Well, here is where the 'significance' of a proposal for cultural universals runs into the culture versus nature question—the second component of the significance issue (beyond the explanatory fruitfulness of universals for a particular cultural domain like language). For if there are certain facts about all cultures, then why aren't these facts simply non-cultural 'physical' facts about the natural creatures which are humans. All humans breathe air, digest food, seek shelter, etc. But those commonalities are due to the 'physical' (biological, physiological, neurological) design of humans. In short, humans have to breathe, eat, and sleep 'by nature'. Even if it is somewhat metaphorical to say so, it is still important to say that what 'nature herself' causes is not caused by individual humans or groups of them. Culture is what is not 'nature caused', but what is 'human made'. This raises a peculiar distinction for the line of thought just sketched about innate knowledge of linguistic universals. For it would appear that vocabulary would be, then, a good candidate for being a cultural phenomenon, whereas underlying facts and rules known—especially, innate rules known—would be non-cultural since biologically determined and 'caused by nature', so to speak. To me, this seems as absurd as the proposal about accessibility to consciousness advocated by Nagel [23] [24] and Searle [36] to the effect that abstract rules that are not even potentially accessible for introspection are part of the 'physical' machinery of the brain, whereas those other rules (pluralization?) that are accessible to introspection are still 'mental' via their potential occurrence in consciousness. Aside from the unclarity of the concept of 'physical' operating here (as Chomsky insists), the inaccessible rules have as much to do with what's known (internalized) about the language in question as the supposedly accessible ones and it is rather implausible to say that one sort of knowledge possession is 'mental' and the other 'physical' (ringing in the old mind-body problem at exactly this point), thereby distinguishing the two as significantly different sorts of knowledge. (Is the 'mental' sort better, more 'knowledge-like' and so the 'physical' sort is just mechanical information processing of lesser importance in the overall account of cognition?)

So, the second challenge to the significance of any proposal for cultural universals concerns reducibility to non-cultural facts. If a proposed universal is only to be considered universal because it's a biological or other 'physical' requirement for existence (e.g., eating), then the so-called universal is just not a cultural one. The basic idea is that what is cultural is arbitrary, conventional, humanly devised or created, variable, and not absolutely dependent on biological or other 'physical' laws or facts. In short, cultural phenomena (activities and practices) could have been otherwise. They are contingent and arbitrary (even if sometimes vital to humans—e.g., medical practice) [4]. For example, no humans had to build or use sailboats. Using sailboats is a human, cultural innovation. You might think that such technology is not completely arbitrary and conventional (since facts of navigation and physics constrain the activity) and thus is not fully cultural. I won't discuss this here, but below I will suggest a new way of viewing the arbitrariness and conventionality of cultural entities (practices, and activities) wherein the conventionality is extracted from certain tangles with contingency, so that genuine arbitrariness and conventionality of cultures does not mislead us in understanding how and what cultural facts and phenomena are.

Well, what's the answer? If a proposed cultural universal really does apply to all humans (and/or all groups of them), why isn't it automatically a biological necessity [5]? The answer I propose is that the universal in question retains the elements of conventionality, arbitrariness, variability, creativity, and contingency—which I label the 'cultural-character' (vs. 'natural-character')—which the nonuniversal aspects of the same cultural phenomena possesses. No one would suggest that the printing press or sailing are culturally universal, but aspects of language, music, religious rituals, buying-and-selling, and many other phenomena do give rise to proposals for cultural universals. Language just is, in particular, the paradigm of what is meant by 'cultural phenomena'—especially, its various special uses (in poetry, drama, law, politics, etc.). Any language's ordinary vocabulary—not to mention those of highly prized sub-cultural areas such as science, law, music, etc.—is paradigmatically arbitrary and conventional. No word has to mean what it does, or be used in the way it is. Language components and uses are simply not what they are 'by nature'. Human intervention—often radical creativity—is required. Nature didn't make the words mean and do what they do, humans did! But then the answer to alleged reducibility to underlying biology for proposed linguistic universals consists in holding that the cultural-character of the linguistic universals is what makes them non-reducible. Universality is no argument for a phenomenon being non-cultural, if the phenomenon retains its cultural-character in some way or other. Now it may not always seem clear that an alleged linguistic universal does retain it, e.g., the universal of every natural language containing nouns and verbs. Maybe that's just a logical necessity. So, where is its cultural-character? One answer comes from applying ordinary logic.

In formalizing English in the terms of the first order predicate calculus, nouns and verbs both are 'regimented' (in Quine's phrase) as logical predicates. With this in mind, someone might propose that since every common noun and verb is a logical predicate and since such predicates are more like verbs than nouns, this shows that the supposed universal difference between nouns and verbs is simply false or vacuous. Now I do not believe that in the end this approach to collapsing nouns and verbs to one syntactic category will hold up. However, seeing that it is a possible proposal—i.e., natural languages might (for all we know) have had no important distinction between nouns and verbs in grammatical structure—shows that there does exist an air of conventionality about the ubiquity of nouns and verbs in natural languages. There needn't have been nouns and verbs universally, since natural languages might have been constructed like the predicate calculus is, with no such distinction. So, even though it is no superficial choice anyone ever consciously made (vs. such conventions as using the color red in traffic lights to indicate stopping), still the cultural-character of conventionality is appropriate to apply in some collective sense. Humans (though no particular human) adopted the kind of languages that have nouns and verbs prominently, instead of adopting languages that do not (like the predicate calculus).

In the end it is the cultural-character of linguistic universals that keeps them from being reduced to non-cultural underlying biological phenomena [6]. This character may seem rather remote when the properties in question are very abstract. But it is there, I contend. Analogously, I contend that any proposed cultural universal has to resist the charge of insignificance due to reducibility to biological (or other) fact. You might compare refined choices of what to eat and how to eat it (surely cultural) with the basic biological determination of humans to eat various plants and other animals. Culturally, we choose to cook or not some of our seafood (and flavor it, etc.), but biologically we do not choose or not whether to try to eat sand. It's not an edible for us biologically. So, the universal rejection of sand as a potential nutrient is not a cultural universal. Similarly, the universal acceptance of at least some plants and animal-flesh as nutrients is not a cultural universal. There is nothing of the cultural-character about it. But deep-frying definitely has the flavor [sic] of a cultural-character. And if deep-frying were found everywhere (and were not otherwise trivial, which would be my guess), then deep-frying would be a genuine (gustatory) cultural universal.

MUSICAL UNIVERSALS

Maybe linguistics is not a convincing source for empirical evidence for cultural universals, because the presence of natural language in one form or another is so pervasive across human cultures that it seems to be a biological necessity—not just a cultural matter. This universal talk in every human society seems as 'natural' as breathing, eating, and sleeping. The cultural-character of natural language as a general phenomenon disappears when we stand back, especially when we 'stand back abstractly' so that the alleged linguistic universals can be appreciated across the varying languages. I believe this complaint can be persuasively answered just using the facts and arguments introduced above. However, instead of answering that way, I will move to another example—music. Lerdahl & Jackendoff have shown that there are at least 52 musical universals of ordinary tonal music [22]. I know I must immediately answer the superficial objection that those can't be cultural universals because so-called 'ordinary tonal music' is just music of the Western World in the Modern Era and so is obviously non-universal. The answer has two parts. First, and less importantly, 'ordinary tonal music' is not limited as just stated. It is found throughout the world and it is hard to say there is a culture today in which it is entirely unknown. Still, of course, examples where such tonal music does (or did) not unequivocally occur will be offered. So secondly, and very importantly, the universals Lerdahl & Jackendoff point out are not features which every piece of music exemplifies. Rather, such claims are often hypothetical; viz., if performers or listeners of music in any culture encounter or perform tonal music, then the structure of their musical comprehension will follow exactly the universal principles that Lerdahl & Jackendoff specify. Many linguistic universals are like this too, hypothetical, which by no means impugns them as universals. Of course, increasing generality of hypothetical form of linguistic universals must eventually be limited, since if we are too permissive, facts of particular grammars (such as that 'bachelors' is a count noun in English) can be trivially deemed universal simply by appending suitable conditional clauses (e.g., as in 'If a language has a word that sounds like 'bachelor' does in English and means what it does in English, then that word is a count noun'—certainly true of any language you pick, though only non-vacuously true of English). But non-trivial hypothetical universals can be easily be imagined—e.g., that if a language has inflection for number (singulars and plurals) in its nouns, then it has noun-verb agreement in number (which is certainly true of several languages and looks like a general proposition that is significantly true of every language (though vacuously of some, if there is a language that does not inflect its nouns for number)).

Well, if there are at least 52 musical truths applying across every culture, are they significant? Significance above consisted of two requirements: first, being explanatory (not failing to play a role in any explanation of the nature, genesis, development, maintenance, or destruction of the culture); and second, not reducing to mere biology. On the first aspect of significance, explanatoriness, I will have to remain silent about Lerdahl & Jackendoff's universals. For I just don't know (and no one else knows either) whether their universals will be explanatorily vacuous. However, I invite the reader to examine their proposals. They are proposals for universals of music comprehension in four areas—viz., grouping structure (or 'phrasing'), metrical structure (the 'beat'), time-span reduction (a more difficult temporal feature of tonal music concerned with the hierarchical structure of melodies [7]), and prolongation reduction (patterns of tension and relaxation that defy brief characterization). It is very clear that Lerdahl & Jackendoff have not offered an exhaustive list of universals of tonal music. For their 52 universals in four areas only cover the temporal and/or dynamic side of tonal music. One might have thought that there were only two temporally interacting features of tonal music, the beat and phrasings. But the other two domains concern patterns through time which are temporal hierarchies as well. All this is quite short of discussing actual melodies determined by specific pitches, the area which Lerdahl & Jackendoff's results set the stage for. If tonal music were thought of as parallel to one language (say, to French), then the universals specified by L & J would not be persuasive instances of innate knowledge. For in the parallel case of learning French, it is not plausible to propose that all of its principles are innately known in order to make acquisition of it possible. So, tonal music (knowing, learning and using of it) has to be taken to parallel general natural-language competence itself (so that postulating innate knowledge of its universals genuinely contributes to solving the poverty-of-the-stimulus challenge). I myself would defend this interpretation, taking tonal music to be music in general, where apparent alternatives to it (in "oriental" scales, or 12-tone or semi-tone music) are regarded as specific modifications or departures from it. Thus, what would correspond to particular natural languages would be particular styles, genres, (even) atonal departures, formats, instrumentations, etc.

of tonal music. When you look at the persuasive details of these musical universals, it is hard to imagine that they will be explanatorily vacuous. And Lerdahl & Jackendoff's interesting exploration (Chapter 12) of possible relations between their 'preference' rules and phenomena in other areas of human cognition (Gestalt phenomena in vision, quantifier-scope interpretation, pragmatic conversational principles, and semantic classification procedures) certainly makes it appear that their universals won't turn out to be explanatorily fruitless or trivial [8].

The second aspect of significance is the non-reducibility to mere biology. Do the universals Lerdahl & Jackendoff propose reduce to noncultural universals because they have no (little, or contaminated) cultural-character [9]. Well, the same kinds of objections that could be raised about the cultural-character of linguistic universals can be raised about the cultural-character of Lerdahl & Jackendoff's musical universals. These universal rules of music structure might be merely the results of psycho-physiological facts (laws, principles, and information processing structures of the brain) to which no cultural-character attaches at all. For structures and operations of the brain neurologically described certainly won't turn out to have significant features of conventionality, arbitrariness, being non-nature-caused (human invented), and contingent.

My response is that if you keep in mind what the 'game' is here—viz., pondering the mere existence plus possible significance of genuinely cultural universals—then even though these reductions to underlying biological fact and phenomena seem to be possible, still language and surely music have been thought to be among the primary examples of what cultural phenomena are. So, if we pursue the reducibility of the proposed linguistic and musical universals to mere biological phenomena, we must eventually land in the position of saying that some features of language and music possession, acquisition, and use are cultural (because they have the cultural-characters—e.g., sailing-vocabulary, linguistically, and waltz-time, musically) and some others are not (because though appearing similar to less abstract rules of particular languages or musical styles, they are actually fixed biological necessities). For myself, this extension of a division between cultural and non-cultural aspects of language to the case of music seems to be a mistake. I would guess that some other way of explicating the cultural-character of both language and music (cf. Section *Essentialism and Natural Cultures* below) will leave language and music just where they traditionally have been while at the same time permitting that cultural universals exist in each area which do not reduce to noncultural, underlying biological phenomena or facts. In this brief excursion into musical universals, I have not mentioned the background motivation which Lerdahl & Jackendoff explicitly adopt. Unsurprisingly, it is (again) the 'poverty of the stimulus'—viz., that the music-learner to acquire the musical knowledge he does (his cognitive musical capacity for hearing and producing tonal music) faces an informationally daunting task (to acquire a large amount of musical information quickly and without much experience, training, or learning). So, innate knowledge of musical universals is proposed to make the acquisition of tonal music competence possible. The reason ordinary humans can succeed at quickly and easily becoming tonally-music competent is that they already know many features of what tonal music is—innately. The original stimulation for Lerdahl & Jackendoff was, of course, Leonard Bernstein's Norton Lectures [15].

Finally, in paralleling music to language, someone might say that granted there are many different types and styles of music found throughout the world, but when we just consider the mere occurrence of some sort of music or other (with no specific details), that is not a suitably 'cultural' matter to consider. For (they might go on) humans are just that way by nature. They breathe, sleep, eat, talk, and (now!) sing or otherwise make music. Well, maybe. But to say this suggests that the same thing will be said of every cultural domain for which universals are proposed. And then this kind of argument for reduction to biology becomes akin to (or just is) a refusal to consider the distinction between two kinds of biologically supported phenomena, cultural and non-cultural (recall note 6 above).

NATURAL CULTURES AND KNOWLEDGE

To many observers, the fact that a capacity for some activity is learned or acquired is sufficient for concluding that it is not a 'natural', innate, and/or instinctual one, and so is clearly cultural. Knowing how to drive a car, or even ride a bicycle, is like this—wholly acquired, and so a matter of culture, not nature. This view quite plausibly leads to the idea that if an activity is even partly innate ('natural' or instinctual), then either it is not cultural (but is 'natural' like sleeping and eating) or the part or aspect of it that is innate is not cultural. Even though there are some human activities that are uncontroversially cultural in not depending on any domain specific innate knowledge—e.g., fly-fishing, driving automobiles—still very many cultural matters are mixed. I would insist that language and music are among these (as, indeed are mathematics, morals, and much else). Is it plausible to divide any such mixed activity (knowledge for which seemingly can't be acquired unless some of it is innate) into two parts—a 'learned' part that is legitimately deemed 'cultural', and an unlearned (though still known) part which is deemed non-cultural and thereby 'natural'? No (recall the above discussion on vocabulary vs. the underlying facts and rules).

Indeed, the existence of mixed cultural capacities (part innate, part acquired) is what permits all cultural activities to be plausibly classified as 'natural' in the broader sense. That is, among everything that humans "naturally" do, some are cultural and some are not (cf. note 6). I judge that all these cultural entities and activities crucially involve acquiring genuine knowledge ('knowing-that's, as well as 'knowing-how's). If there are cases of human states or activities that require knowledge of a type that is wholly innate, then perhaps those phenomena should not be deemed culture. (Are vision and hand-eye coordination among these? Or, should all the innate capacities for these, which mature very early and quickly in humans, be considered not to be knowledge at all—having no components that are knowledge-like?) The two cases I have discussed vis-à-vis significance of cultural universals—language and music—certainly do involve the possession, acquisition, and use of much knowledge. The apparent fact that some of it is innate is no good reason for concluding that language and music are not genuinely cultural matters—that language and music are non-cultural (merely natural). Similarly for any specific part or aspect of such activities—such as linguistic or musical universals that are innately known. They are genuinely cultural too, because they are intimately involved in the mixed (innate and acquired) cultural phenomena of language-use and musical- performance / recognition.

ESSENTIALISM AND NATURAL CULTURES

I propose that the tension between the kinds of universals proposed above and their possible reducibility to non-cultural biological necessities can be resolved through a better understanding of the alethic modalities involved. One of the components of the required cultural-character is contingency. But the universals proposed (linguistic and musical) seem to be necessities, at the very least in the sense of biologically necessary. That is, natural languages have to be such that the rules and principles innately known (the content of linguistic universals) are characteristics of every one of them. Otherwise, we couldn't learn them. A similar picture results for musical universals. Tonal music has to be the way it is due to our innate musical dispositions (dispositions which are such as to be correctly labeled 'knowledge'). Where's the contingency in this? If you say it resides in the contingency of the underlying biological laws, that is just insufficient. For then there would be no difference contingency-wise between sleeping and eating, on the one hand (necessitated by contingent biological laws), and music on the other.

The answer to these questions rests in revising the nature and role of contingency as a component of cultural-character. Surprisingly, the revision consists in applying an essentialism wherein the contingency of cultural universals seems to recede further. However, the true locus for contingency eventually reappears (in the choices and free actions of members of a culture). I will briefly explain this by applying the variety of essentialism recently created by Kripke and Putnam [10]. I apply it to linguistic expressions themselves, an application Kripke and Putnam did not carry out. Even though they stated their views via analysis of linguistic expressions, it was analysis of how linguistic expressions are used to refer to things and phenomena in the extra-linguistic world. My extension of the approach applies it to linguistic expressions themselves, as if they were (which I believe they are) simply ordinary things and phenomena in the world as much as water, wind, George Bush, and the moon are.

First, consider ascriptions of analyticity, synonymy, and meaning to linguistic expressions; e.g.,

(1) 'Bachelors are all and only unmarried males' is analytic.

(2) 'Bachelor' is synonymous with 'unmarried male'.

(3) One meaning of 'bachelor' is what is expressed in French by 'célibataire'.

Concentrating on (2) as a representative of these, it is plausible as a starting point to say that (2) is contingent. 'Bachelor' did not have to turn out to mean what it does, and the concept of bachelors (the kind of thing bachelors are, certainly a cultural role if anything is) did not have to turn out to be expressible by 'bachelor'. Application of essentialism to words themselves (considering the word 'water' and its essential properties rather than the stuff water and its essential properties, like being HOH) produces a different picture [11].

First, what are essential properties of the word 'bachelor'? At the very least, they are phonological. If 'bachelor' was pronounced with an initial unvoiced bilablal stop, it would sound like 'pachelor'. But that would be a different word, not exactly the same word just accidentally modified. Changing such a sound is changing something essential (thus destroying the word, even if a very similar word takes its place). Other aspects of the sound of 'bachelor' are accidental; e.g., the speed with which it is pronounced, or the pitch (said in a high voice, or a deep one, it is still the same word). The view of words in which the sounds alone are essential,

I have labeled 'phonologism' [30]. Phonologism on words is indefensible, though I won't spell that out here (cf. ibid.). For syntactic properties are also essential to word identity, and so are semantic ones—literal meanings and semantic role (cf. ibid.).

The resulting view of word identity then is that a word is a sound-form-meaning complex wherein all three kinds of properties are essential. That means that (2) turns out to be a necessary truth, rather than a contingent one! In short, not only would 'bachelor' not be the word it is, if it were pronounced differently (though dialectical variations can often be accommodated), it would also be a different word if it were a verb or a preposition or if it meant something else (e.g., meant what 'mortgage' now means in English). The essentiality of semantic characteristics to a word is what is absolutely new in this approach. For the sake of brevity, the only defense I shall give of this addition will be its payoff below.

So, (2)'s truth is necessary because it concerns the essential nature of the words and phrases referred to in it—not because of the mere meaning of the words in it. Inattention to how use-mention quotes are used can hamper understanding here. The problem is how to understand such names. On my approach, quotation-mark names are taken to be genuine proper names—viz., analogous to proper nouns with no literal linguistic meaning attached. (2) does seem to be analytic (and so not simply necessary-if-true-at-all, but necessary-because-analytic) when the quotation mark names are taken to be definite descriptions. Thus, if the expression which is the grammatical subject of (2) is understood to be what is expressed by "The word in English pronounced like 'bachelor' is, which is a count noun and means what 'unmarried male' means", and if the grammatical object of the final preposition in it were analogously interpreted, then (2) would be analytic (and still not contingent, as it first seemed). Such a version of (2) could be:

(2)d The word which is pronounced like 'bachelor' is, and which has the meaning that "unmarried male" does, means the same thing as (is synonymous with) the phrase which is pronounced like 'unmarried male' is, and which has its meaning.

And (2)d is true solely because of the meanings of the words in it. That is, (2) as (2)d would be necessarily true for the same reason that "Bachelors are all and only unmarried males" is—because analytic.

So, if you take quotation-marks names to be abbreviated definite descriptions, (2) won't be necessary because of the essential nature of the words referred to in it, but because of the meanings of the words used in the expressions used to refer to those words. To grasp my application of

Kripke and Putnam here, replace the quotation mark names with stipulated names for the word and phrase in question—say, the letter 'b' as the name for the word 'bachelor' (replacing the expression 'bachelor' in (2)) and the letters 'um' as the name for the phrase 'unmarried male' (replacing the expression 'unmarried male'). Then, the following is easier to understand as expressing a truth that is necessary because of the essential nature of the referents:

(2)n b is synonymous with um.

If the nature of what 'b' refers to in (2)n is such that it has to be synonymous with what 'um' refers to, then (2)n is a necessary truth—but not necessary because it's analytic. So, if the semantic content of b is essential to it (rejecting phonologism, as well as phonologism + syntacticism, as an account of word identity and accepting the full package of sound, form, and meaning as what an expression essentially is), then (2)n can be regarded not only as necessary, but also empirical! On my approach, quotation marks names are genuine proper names, so (2)n is merely a statement that says exactly the same thing that (2) does (but with two different proper nouns). This means that (2)—when interpreted in a way that Kripke-Putnam essentialism could be plausibly applied to it—is not contingent, but still is empirical. For all we know English could have developed differently, but given how it did develop the word 'bachelor' has to mean what it does and we discover this empirically (via internalization in ordinary language-acquisition of English, and reflectively in scholarly and/or scientific inquiry into the nature of English). So, to interpret the quotation-mark names in (2) as definite descriptions makes (2) into an analytic truth, rather than its apparent (to start with) contingency. (The only way I can see to reasonably argue for the contingency of (2), with quotation-mark names taken as definite descriptions, would be to adopt phonologism or phonologism + syntacticsm on word identity.) A similar thing would happen to (1) if the quotation-mark name of a sentence in it were taken as a definite description (containing in the description expression of what 'bachelor' means). For then (1) so modified would be analytic, rather than its apparent contingency. On my reinterpretations the apparently contingent (1)-(3) all turn out to be necessary, but not necessary because analytic. Also, I harp on semantic properties, but ascription of phonological and syntactic properties would be just as empirically necessary—e.g., the empirical necessities 'b begins with a bilabial stop', and 'um is a noun phrase'.

However, this seems to lead to the conclusion that the alleged contingency of arbitrary and conventional (humanly created, not nature dictated) word-meaning relationships is bogus. For meanings are no longer contingently possessed. But there is a saving grace to this loss of contingency. For the semantic nature of such words is discovered empirically. We discover that water is HOH empirically (though it turns out to be necessarily so). Similarly, we discover what 'bachelor' means empirically (even though it is equally necessary). Does this mean that words too like water and trees are natural kinds? What I would say is 'perhaps' (though earlier (1989) I said 'yes'). For the Kripke-Putnam approach to natural kinds is rather linguistical, and I would think that, in the end, there must be more to natural kinds than these linguistic features of natural kind terms suggest (e.g., that natural kinds are nature at the 'joints' and are mentioned in fundamental scientific laws). However, if we restrict ourselves to just the Kripke-Putnam notion of natural kinds—their being what the natural kind terms refer to (cf. Peterson [29] [31] [32])—then words and phrases themselves turn out to be natural kinds too [12]!

I will be very brief in explaining the lesson of this approach for the cultural-character of linguistic and musical facts and phenomena. A relevant criticism of my view of linguistic expressions (essential phonological, syntactic, and semantic properties) is that, semantically, it resuscitates Saussure's view of the 'sign' as both 'signifier' (= sound) and 'signified' (= meaning or concept, not referent). The criticism will be that just as Saussure's views on phonology and syntax have been superseded by linguistic advances in the 20th century, so similarly his semantic views have been used and replaced. For various approaches to developing a natural language semantics without the Saussurean 'signified' (qua concept and/or meaning) have been developed—from (i) a Quinean view in which there are no such mentalistic or abstract meanings but only 'dispositions to respond' (which eventually produce impermeable semantic indeterminacy, with resulting ontological relativity) to (ii) varieties of 'use' theories in which linguistic meanings are explicable as a function of speaker intentions and community conventions (Grice-Searle-Lewis-Bennett) to (iii) postmodernistic despair in postulating any 'signifieds' or replacements so that signs are reduced to signifiers capriciously used to enforce or rebel against (intra- and inter-) cultural relationships of social power. My response is brief. None of these alternatives (nor many other rivals developed, such as Montague Semantics) have succeeded in replacing Saussurean semantics; i.e., his phonological and syntactic insights have been used and then improved on, but not so his semantics. For 'signifieds' remain; cf. Peterson [25] [26] [27] [28] [33]. Also, see the 'intensionalisms' of Katz (e.g., [19]) and Bealer (e.g., [14]). What I have offered herein is the appropriate attention to the logical details of the alethic modalities that any Saussurean semantics deserves (wherein signs possess what they signify essentially). Take a very simple tonal melody like that of the song 'Happy Birthday'. It ends on the tonic, like orthodox tonal melodies do, but it begins on the dominant. Although Lerdahl & Jackendoff did not propose any universals about melodies, pretend for the sake of discussion that a good hypothesis for a melodically germane musical universal is as follows: prefer (even strongly) tonal melodies which end on the tonic. Does this mean that the fact that the composer of Happy Birthday made the tune so that its ending on the tonic is necessary truth (though only empirically discovered)? No, certainly not. If Happy Birthday is to be a tonal melody, then that feature is strongly preferred. But given exactly what Happy Birthday already is, it has to end on the tonic (since inter alia it was orthodoxly composed). Now if you say that the tune could have been composed so that it ended on the dominant like it begins, or someone could sing it that way, the proper response is, 'No, it could not, for if it had been composed that way it wouldn't be our Happy Birthday, but some other tune.' Further, if you sing it that much off (small departures are permitted in singing, but not huge ones)—so that you end on the dominant, then you have not sung Happy Birthday!

The main way contingency comes into cultural matters is in the 'doing' or 'having' of them. Recall Kripke's idea about baptism with proper name. If Harry Truman says (in the '20s, appropriately) 'This baby is [=] Margaret', then he creates a rigid designator in an identity statement that is *a priori* contingent—rather than empirically necessary. The point is that when an object is so baptized with a proper name (the proper noun thereby becoming its name), the baptism did not have to happen. It is a contingent fact that 'Margaret' became the name of HST's only child. But it was known *a priori* in that, first, it was not an empirically discovered fact by HST himself and, second, it came about through HST's decision. (He knows his decisions because he makes them, not because he or anyone else empirically observes them. I am often inclined to call Kripke's contingent *a priori* truths 'contingent decisions, knowable *a priori*'.) Of course, for us to find out later who Margaret Truman is, we have to find out empirical facts about her—contingent ones, or even necessary ones (such as that she is the daughter of HST or that Margaret = Paula Hume where 'Paula Hume' is another rigid designator for her and we 'know who' Paula Hume is).

Think of the composer of Happy Birthday as doing something in his composing rather like HST does in baptizing his daughter. By *a priori* decision, the composer makes something true contingently. He makes or invents the melody. Afterwards, he can give it the name in a baptizing not too unlike HST's (though an abstract object, a tune, is named, rather than a concrete one). This is no fundamental problem for, or departure from, the Kripke-Putnam approach, though it requires mentioning something they did not go into—viz., that natural kinds themselves are abstract objects (or universals). Cf. Peterson [31] [32]. But compare the baptism with an act of musical creation. It's contingent that the sounds—in both cases (linguistic and musical)—are 'made' to be what they become. It's a contingent fact that Margaret Truman was named 'Margaret', but when she has been so baptized, then we can use that name in making empirically necessary statements like 'Margaret Truman is the daughter of HST'. Similarly, it's a contingent fact that the string of pitches that constitute the melody of Happy Birthday were ever put together in the way we call 'composing' so that a repeatable tune came into existence (that was eventually labeled 'Happy Birthday'). Similarly, with actually singing the song (or humming the melody), it is definitely contingent when and/or whether (when you try) the melody is actually performed. So, its creation was definitely contingent, as well as its being put to use. But if it is successfully put to use, then that performance has to begin on the dominant and end on the tonic, among several other things. For that is the 'nature' of the tune. In sum, its existence and nature is, in one way, quite contingent and accidental (for the composer of it did not have to compose it or anything at all and did not have to compose it the way he did), but given the composition in existence (and already so labeled) Happy Birthday has to have, at the very least, the metric and pitch characteristics it does have. Happy Birthday is that tune. Margaret Truman is that daughter of HST. Both by necessity, even though either one (person or tune) might not have existed in the first place.

The moral of this approach for describing and explaining cultures, such as with the use of empirically discovered universals, is that the cultural-character of cultural phenomena pertains to the creation of cultural entities and the performance or use of them. For example, it is contingent exactly where the Mona Lisa (a concrete particular object) hangs right now (as well as when it was created), but it is not contingent that it has certain of its visually detectable properties (and probably many other material properties). Being humanly made entities does not require that their properties are all contingent or accidental. Their essential features are necessary to them. And this need not impugn the other aspects of their cultural-character. For the way in which Happy Birthday (that melody) is conventional and arbitrary is not that its essential features aren't really necessary to it. Rather, it is conventional and arbitrary because (i) we didn't have to have that melody at all, (ii) we didn't have to have it exactly as it is (the composer didn't have to put the pitches together that way), (iii) it didn't have to be labeled 'Happy Birthday', and (iv) we are not compelled to perform it (or even think it) at any particular time, or ever. All of these sources of variation about the creation and use of the melody produce contingencies about it. But the melody itself, as well as the features that any series of pitches has which is a tonal melody, are necessities—even empirical necessities. My own view is (like Kripke's) that these necessities are not something less than full-fledged necessities—say, 'physically' necessary (or necessary via law which holds contingently), though still logically and/or mathematically contingent. My inclination is to say that even though such necessities aren't disguised logical or mathematical truths, still they support relevant statements expressing them which are true in every possible world. Some challenges for orthodox modal logic, or its use, certainly develop in that necessary truths arise which are not necessarily necessary, thus restricting our 'logic' to systems weaker than S4 (which are undesirable due to other shortcomings).

So, the seeming loss of cultural-character by cultural universals (linguistic ones, and musical ones) is no argument against their significance or irreducibility. For with the innovative extension of essentialism I have introduced, it turns out that seeking arbitrariness and conventionality, like seeking contingency, in the nature of cultural entities themselves is a mistake. The locus of contingency—and thereby with it, the conventionality, arbitrariness, and human (vs. nature's) creativity as well—has been shifted. It is shifted from the cultural entities in themselves (what they are and how we comprehend them) to the circumstances and facts of their creation and use.

Finally, on culture versus nature. To be a human, I have to, by biological necessity, breathe air, obtain nourishment, and sleep. Laws of my nature will be obeyed, on pain of my destruction. If I try to control my breathing, for example, such as by ceasing to breathe, then 'nature herself' will take control. For after a short time, I will pass out and then begin breathing again (or die). Now the laws (via relevant cultural universals) about culture are no less necessary than those about my respiration. However, there is an optionality about the creation and use of cultural objects and processes. I do not have to speak at all, or ever sing (or otherwise create or comprehend music). It may be hard to not speak, or not sing, but it is possible in exactly the way it is simply not possible for me to choose to quit breathing (or even modify it much). Nature controls breathing (not just that we can't create the process and it has been created for us, but also how we carry it out—non-agentially, I would say). But when cultural matters are involved, we do have some control. At bottom, it comes down to that—free action, necessarily guided by relevant knowledge and belief. And, of course, those determinists—like so many behaviorists, pragmatists, empiricists, and even Postmodernists—who can't find free action anywhere, are bound to be the least convinced by my distinguishing culture from nature in this way.

How does this Kripkicizing of cultural entities and activities help to avoid reducing culture to nature (and/or cultural universals to biological necessities)? First, the main argument for irreducibility in the first four Sections above was that the cognitivity requirement for cultural phenomena (that cultural participants acquire knowledge to put to use in their activities) preserves their culturalness (roughly, sleeping and digesting don't require any knowledge acquisition, but marrying and whistling Dixie do). But the cultural character of these phenomena had somehow to be honored. I have shown how to honor it in a way which coheres with the cognitivity of culture [13]. *Philip L. Peterson*

Syracuse University,
Syracuse, New York,
USA
e-mail: plpeters@mailbox.syr.edu

REFERENCES

[1] Two 19th century figures are closely associated with these uses of 'culture'—E.B. Tylor for the anthropological and Matthew Arnold for the humanistic. Notice, however, that the appellations for the two concepts are my coinages and are not terminology of typical discussions of culture (more's the pity). Cf. note 3 of [29] for an example of a typical tentative move to acknowledge the two concepts as fundamental, followed by an immediate conflation of them. Also, English dictionaries contain other senses for 'culture' such as 'result of cultivating micro-organisms for scientific study'. These are irrelevant to the two senses I considered to be fundamental—except etymologically of course.

[2] The ideas and terminology of Peterson [29] permit a more accurate description of C.P. Snow's [37] discussion of 'the two cultures'. Snow was describing a difference between ways of thinking and acting largely due to the difference between science (and engineering), on the one hand, and humanities on the other—a distinction which would be more accurately described with the two uses of the word 'culture' carefully distinguished, though his two cultures are not direct instances of my distinction. However, he was using 'culture' in just the anthropological sense to do it. Cf. notes 2 and 4, [29].

[3] One question that comes immediately to mind is, if the Postmodernists are right, is it a cultural universal that politics controls all action, explanation, and evaluation in science, art, literature, and life? Here's another: aren't Postmodernists explicitly engaged in equivocating on the two senses of 'culture' of

Peterson [29]—even to such an extent that if the two senses were honored their whole approach would collapse?

[4] Of course, contingency alone would not prove anything, since the facts and laws of biology, or other 'natural' phenomena, are ordinarily taken to be contingent as well—i.e., not necessary in the traditional sense in which facts of logic and mathematics are paradigmatically necessary and facts of the 'natural world' are paradigmatically contingent. I return to this topic below in Section *Essentialism and Natural Cultures*, advancing an alternative view.

[5] Even if we do not adopt metaphysical essentialism, as I will below, still facts or phenomena dictated by the laws of biology, chemistry, and physics can be considered 'necessities' while not supporting propositions true in every possible world. For they can be deemed 'physical' necessities which are still logical-mathematical contingencies. What's true by laws 'has' to be, no matter what the modal status of the laws themselves—in this case contingent.

[6] Care must be taken to realize what I am saying here. In a broader sense of 'biological', all cultural phenomena are obviously biological via being supported by biological systems which are humans (not to mention everything else biological and physical in the human ecological niche). (I certainly do not believe that cultures have any origins outside the 'natural world'—in the 'supernatural', sacred, or otherworldly 'spiritual'—or even outside our immediate planetary zone from science-fictional aliens or places.) My aim here is to distinguish those aspects of biological phenomena broadly-speaking that are properly called 'cultural' from those that are not (and are merely biological, in the narrower sense).

[7] Think of the succession of ever simpler melodies one might generate out of a given tune by, first, omitting the pitches which seem absolutely the least important to the tune (pitches which when omitted still easily permit the melody to be detected), and then moving on to generate a third melody out of the slightly simplified first by repeating the process, and then moving on in the same way to generate a fourth, fifth, etc. simplification for as far as you could go until one or two dominant pitches result (pitches which do not permit recognition of the original melody, but which do somehow characterize the melody at the highest level of abstraction). Now if this process is carried out in the way intended a hierarchy will be developed of the pitches in the original melody, with those pitches which survive longer 'dominating' in the hierarchy those that don't. This is not the whole picture with time-span reduction (a kind of recognition of hierarchical structure in strings of pitches, where the structure is not ordinary beat or ordinary phrasing—both of which cut across these hierarchies, as well as being hierarchical themselves). For the subordinate links between dominating and dominated (elaborative) pitches must also be acknowledged. However, this is the best I can do briefly.

[8] Make this comparison: select one of L & J's universals—say, GPR 5 (Symme-

try): prefer grouping analyses that most closely approach the ideal subdivision of groups into two parts of equal length—and compare it with what surely would be a trivial (explanatorily fruitless) cultural universal, if it were a cultural universal (which I am sure it is not), viz., the practice of deep-frying seafood. I can't now see how anything important or illuminating will turn out whether deep-frying of seafood is universal or not. It seems to me to be a curiosity (if universal, which I doubt). Of course, very strange things happen in science and scholarship. So, maybe some Postmodernist will soon show how such a 'marginal' or simple-minded aspect of cooking can be shown to make a great contribution to our understanding of cultures or sub-cultures. On the other hand, GPR 5 does seem to me at least to have some potential for contributing to interesting explanatory tasks, both within direct anthropological or sociological studies of cultures and within (for example) cognitive psychology.
[9] Remember that every genuine cultural phenomena, particular or universal, must have biological support or grounding. The question about reducibility is whether any biologically grounded universal phenomena are non-trivially deemed 'cultural'. The only way I can see so far that irreducibility could be defended is by requiring genuine cultural-character—viz., some degree of application of the usual idea of cultural phenomena being somehow conventional, arbitrary, contingent, and human-made rather than nature-caused.
[10] Cf. Kripke [20] and Putnam [35]. Also, what I shall present as Kripke-Putnam essentialism is, of course, my own understanding of what it is. I have given some details before of what this is (my understanding); cf. Peterson [29] [31] [32], and forthcoming.
[11] Statements about language like (1)-(3) parallel statements about cultural entities like

(4) Both wrenches and pliers are made to grip things tightly, better than you can with your bare hands.

(l)-(3) are meta-language statements and (4) is a meta-tool statement. A statement like

(5) Bachelors are all and only unmarried males.

(which is deemed analytic in the meta-linguistic (1)) does not appear to have parallels in the non-linguistic domain of (say) mechanical sub-culture. In using (5), you use some words (with their meanings) to express something—roughly, applying words to things (complexly). That suggests to me that the tool-world parallel of a statement like (5), which turns out to be true merely for linguistic reasons despite the fact that it is about human males, would be (say) some use of the pliers that resulted in vacuity too. Perhaps, tightening and loosening the pliers simply to reveal (by the action) what the pliers are.
[12] Some of those readers unhappy with the Kripke-Putnam account of reference, meaning, and natural kinds, can (of course) take this result to be the absurdity proving them wrong.

[13] A revised version of this paper is forthcoming in American Philosophical Quarterly under the title 'Do Significant Cultural Universals Exist?'

[14] **Bealer, G.**, "Foundations Without Sets", *American Philosophical Quarterly*, **18**, 1981, pp. 347–353.

[15] **Bernstein, L.**, *The Unanswered Question*, Harvard University Press, 1976.

[16] **Chomsky, N.**, *Some Concepts and Consequences of the Theory of Government and Binding*, MIT Press, 1982.

[17] **Chomsky, N.**, *Knowledge of Language: Its Nature Origin, and Use*, Praeger, 1986, Chapter 2.

[18] **Kaplan, D.** & **Manners, R.**, *Culture Theory*, Prentice Hall, 1972, p. 3.

[19] **Katz, J.** & **Katz, F.**, "Is Necessity the Mother of Intention?", *Philosophical Review*, **86**, 1977, pp. 70–96.

[20] **Kripke, S.**, *Naming and Necessity*, Harvard University Press, 1980.

[21] **Kuspit, D.**, "The Contradictory Character of Postmodernism." In: **Silverman, H.** (ed.), *Postmodernism: Philosophy and the arts*, Routledge, Chapman and Hall, 1989. (Also, in: (**Bender** & **Blocker** (eds.), *Contemporary Philosophy of Art*, Prentice Hall, 1993.)

[22] **Lerdahl, F.** & **Jackendoff, R.**, *A Generative Theory of Tonal Music*, MIT Press, 1983.

[23] **Nagel, T.**, "Linguistics and Philosophy." In: **Hook, S.** (ed.), *Language and Philosophy*, New York University Press, 1969.

[24] **Nagel, T.**, "Review of The Rediscovery of the Mind" (by J. Searle), *New York Review of Books*, March 4, 1993.

[25] **Peterson, P.L.**, *Concepts and Language*, Mouton, 1973.

[26] **Peterson, P.L.**, "Philosophy of Language." *Social Research*, **47**, 4, 1980, pp. 749–774.

[27] **Peterson, P.L.**, "Semantic Indeterminacy and Scientific Underdetermination." *Philosophy of Science*, **51**, 1984, pp. 464–487.

[28] **Peterson, P.L.**, "Revealing Designators and Acquaintance with Universals." *Nous*, **20**, 3, 1986, pp. 291–312.

[29] **Peterson, P.L.**, "The Two 'Cultures'", In: *Philosophie et culture, actes du XVIIe Congres mondial de philosophie* (Montreal 21–27 August 1983). Section 1A, Culture et nature, Volume II, Editions Montmorency, Montreal, 1988, pp. 88–93.

[30] **Peterson, P.L.**, "Logic Knowledge." *The Monist*, **72**, 1, 1989, pp. 78–116.

[31] **Peterson, P.L.**, "Universals and Natural Kinds." Read to the American Philosophical Association Pacific Division Meetings, Los Angeles, California, 29 March 1990.

[32] **Peterson, P.L.**, "Natural Kind Terms are Not Revealing." Read to the Institut fur Philosophie, Universitat Salzburg, 16 May 1991.
[33] **Peterson, P.L.**, "What is Empirical in Mathematics?" *Philosophica Mathematica*, **6**, 1, 1991, pp. 91–109.
[34] **Peterson, P.L.**, "Are Some Propositions Empirically Necessary?" Forthcoming in *Philosophy and Phenomenological Research.*
[35] **Putnam, H.**, "The Meaning of 'Meaning'", In: **Gunderson, K.** (ed.), *Language, Mind, and Knowledge*, University of Minnesota Press, 1975.
[36] **Searle, J.**, *The Rediscovery of Mind*, MIT Press, 1992.
[37] **Snow, C.P.**, *The Two Cultures and the Scientific Revolution*, Cambridge University Press, 1959.

ARNOLD DE LOOF

COMMUNICATION: THE KEY TO UNDERSTANDING THE VERY NATURE OF 'LIFE', 'DEATH' AND THE FORCE DRIVING EVOLUTION

INTRODUCTION

The term 'life' as it is daily used has different meanings in, e.g., 'life activity', 'life span', 'life quality', 'life cycle', 'life as the difference making property between inanimate and animate matter', 'life' returns in a 'sleeping leg', etc. This is one of the reasons which leads to the erroneous impression that 'life' is such an abstract term it cannot possibly be comprehended by the human mind and could not be defined. In this paper I will show how plausible definitions of 'life' (as an activity) and 'death' can be simply derived by logically combining (i) the principles of compartmental organization of the living state, with (ii) the principles of communication and (iii) those of Ilya Prigogine's dissipative systems.

We will come to the conclusion that the terms 'Life' and 'Death' only make sense if they are linked to a 'compartment'. From the observation 'death ensues when a given compartment irreversibly looses its ability to communicate at its highest level of compartmental organization', it follows that 'Life of a given compartment at a given moment is the total sum of all acts of communication performed by this compartment, from its lowest level of compartmental organization to its highest one at that moment'. Living systems cannot do else than evolve and be far from their thermodynamic equilibrium because their communication requires labile gradients. These insights are pondered upon in the present contribution.

CRITERIA FOR AN ACCEPTABLE DEFINITION OF 'LIFE'

According to Schejter and Agassi [1], an adequate definition of 'life' should include: 'Apart from its not being trite and uninformative (circular, to use a traditional term), it should be neither too wide nor too narrow; it should not exclude living things and it should not include dead ones. Furthermore, it should not make biology part-and-parcel of chemistry and physics.' Over the years a variety of approaches has been used to try to define 'Life' but none has as yet been successful [2] [3] [4].

© 1999 *Kluwer Academic Publishers. Printed in The Netherlands.*

DEDUCTION OF THE DEFINITION OF 'DEATH': THE KEY TO OPEN THE 'BLACK BOX' OF 'LIFE'

If one asks the right questions in the right order, and if one answers these questions correctly, then a definition of 'life' and 'death' can be obtained by a rather simple logical deduction as will be shown next.

First question:

What is missing in the following sentences?

- 'There is joy because of birth'.
- 'There is sorrow because of death'.

These sentences are ambiguous because it is not indicated *who* or *what* is born (a baby daughter, a baby horse etc.) or died (grandfather, the cat, our goldfish etc.). This means the concepts 'death' and 'life', in order to make sense, have to be connected to what I refer to, with a general term, as a '*compartment*' or '*system*'.

First observable fact:

'Life' is invariably linked to some 'compartment' or 'system'.

Second question:

What types of living compartments do exist in Nature?

By using only a few simple principles, Nature succeeded in generating numerous types of compartments: cell organelles, cells, tissues, organs, whole organisms, populations, social compartments, food chains, etc.. Figure 1 schematically represents a system, comprising 15 levels, to bring some order in the 'creativity' and evolution of biological compartmentalization.

Second observable fact:

There are different levels of compartmentalization and the majority of compartments contain inside themselves one or several levels of subcompartmentalization

Third question:

At what moment does any living compartment/system cease to be alive? Or: What is the difference between being alive, no longer alive, or not yet being alive? Or: What crucial parameter changes at the very moment 'life' ends and 'death' takes over?

Such questions can be answered by searching for the common denominator in following situations, which correspond to 4 different levels of compartmentalization.

Let us consider some examples at different levels:

A. The multicellular organism level:
Suppose I decapitate a chicken (compartmentalization level 7 in Fig. 1). Is this chicken then dead? One will probably reply that the chicken is indeed dead even if it can continue to move for a while. Suppose now this decapitation is done in a laboratory equipped for cell/organ culture and that I succeed in bringing all cells and tissues of the decapitated chicken into culture and it all works out well: all cells survive. The same question again: Can we still claim the chicken is dead? Even in this situation we will conclude the compartment 'chicken' no longer exists, despite the fact parts of it survive.

B. The eukaryotic cell level (compartmentalization level 2 in Fig. 1):
Eukaryotic cells contain several membrane-limited cell organelles, e.g., the nucleus, mitochondria, chloroplasts (if they photosynthesize), etc. Imagine we have a jar filled with water containing single cell green algae. With a laboratory homogenizer it is possible to homogenize the algal cells in such a way that the limiting membrane of these cells, which is called the plasma membrane, is ruptured but the cell organelles stay intact. Next one can use a technique, called density centrifugation, to collect the different cell organelles. Chloroplasts obtained this way can be injected into a chicken egg. If this egg is illuminated, and if everything goes well, the chloroplasts will multiply and colonize the egg so that it turns green. This behavior of chloroplasts can be explained by the fact that such organelles are supposed to be descendants of bacteria that came to live inside another cell long ago in evolution and continued to behave like bacteria. Thus, the chloroplasts are still alive but the homogenized eukaryotic algal cell-compartment is dead.

C. The population level (level of compartmentalization 13 in Fig. 1):
The same duality is also present at this high level of compartmental organization. Imagine all individual members of a small population of animals get somehow so far separated from each other that they completely and irreversibly loose contact. The individual members are still alive, they can metabolize, grow, cope with entropy, respond to a variety of stimuli and adapt to short term changing conditions in the environment, evidently within certain limits. Despite all of this, we conclude the compartment 'population' no longer exists.

D. The level of the prokaryotic cell or of membrane-limited cell organelles (level of compartmentalization 1 in Fig. 1).
At the primitive end of organizational complexity dual existence is no longer found. After homogenization nothing remains alive. A mitochondrion, a chloroplast or a prokaryote ceases to live after irreversible dis-

ruption of its limiting membrane, e.g., by ultrasonic homogenization. Under proper conditions membrane fragments can still perform some metabolic functions, they are complex, but they are not considered to be alive. Disruption of the plasma membrane ends existing gradient(s) across this membrane.

Figure 1. Levels of compartmentalization and means for generating daughter compartments. The simplest level of compartmentalization is the monomembrane type to which belong the bacteria and membrane-limited cell organelles like mitochondria and chloroplasts (1). The most complicated one which is the one with the largest number of different subcompartments, is the planetary one (15). Different levels of compartmentalization are possible within one and the same individual (1-8). Some compartments are composed of more than one individual belonging to the same species (9-13) or to different species (14-15). For generating additional levels of compartmentalization, several systems are used. A widely used one is the internalization of novel compartments in existing ones: 1. membrane-limited compartments within an outer membrane compartment (the eukaryotic cell type, 2); 2. epithelium-limited compartments within an outer epithelium (6); 3. an organism(s) within an organism (12, 14); 4. a subpopulation(s) within a population (13). Another one is the aggregation of compartments (3,4,5,9 etc.). The communication between the constituent compartments can be intermittent, e.g., through gap junctions between neighbouring cells (4a) or more permanent like in case of cytoplasmic bridges, e.g., in Fungi, or in meroistic insect ovarian follicles or in myoblasts which fused into a muscle fiber. A third system is segmentation (7). A fourth one is the enlargement of a compartment by tools (tool utilization or toolization) (8). A fifth one is the spreading of the genes needed for reproduction over more than one individual (10,11). For segregating a daughter compartment(s) from a parent (reproduction), the variability in systems is rather limited: mitosis, meiosis, systems for asexual reproduction (e.g., 3,6), systems for speciation (13) etc. Evidently, in some systems several mechanisms can be simultaneously operational. Other approaches for categorizing levels of compartmentalization than the one used in this figure are possible. The majority of the levels of compartmentalization depicted in this figure correspond to revolutionary steps in macroevolution. Classical biology mainly deals with the description of the generation and functioning of the different levels of compartmentalization as outlined in this figure. Communication biology focuses on the interplay of all systems for communication (for figure, see [7]).

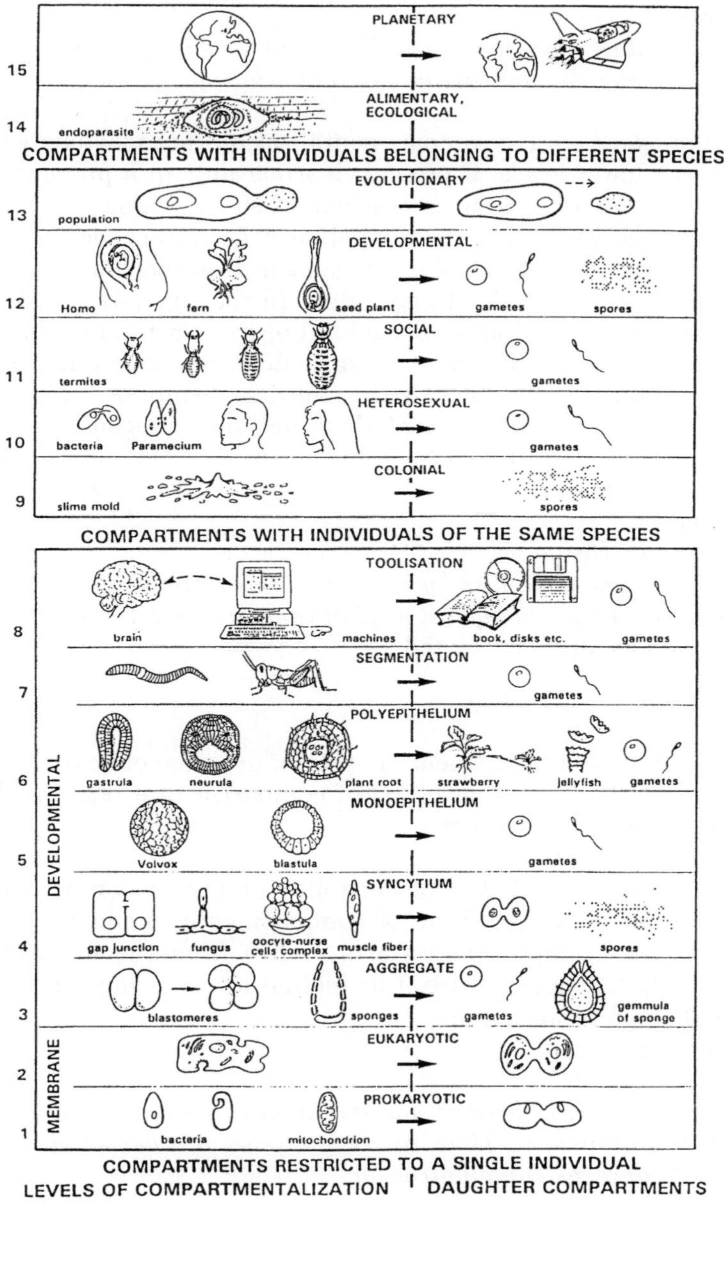

15
PLANETARY
14
endoparasite
ALIMENTARY, ECOLOGICAL
COMPARTMENTS WITH INDIVIDUALS BELONGING TO DIFFERENT SPECIES
13
population
EVOLUTIONARY
12
Homo
fern
seed plant
DEVELOPMENTAL
gametes
spores
11
termites
SOCIAL
gametes
10
bacteria
Paramecium
HETEROSEXUAL
gametes
9
slime mold
COLONIAL
spores
COMPARTMENTS WITH INDIVIDUALS OF THE SAME SPECIES
8
brain
machines
TOOLISATION
book, disks etc.
gametes
7
SEGMENTATION
gametes
6
gastrula
neurula
plant root
POLYEPITHELIUM
strawberry
jellyfish
gametes
5
Volvox
blastula
MONOEPITHELIUM
gametes
4
gap junction
fungus
oocyte-nurse cells complex
muscle fiber
SYNCYTIUM
spores
3
blastomeres
sponges
AGGREGATE
gametes
gemmula of sponge
DEVELOPMENTAL
2
EUKARYOTIC
1
bacteria
mitochondrion
PROKARYOTIC
MEMBRANE
COMPARTMENTS RESTRICTED TO A SINGLE INDIVIDUAL
LEVELS OF COMPARTMENTALIZATION
DAUGHTER COMPARTMENTS

What is the common denominator in all these examples? Which criterion did we use to conclude that 'life' ended?
Third observable fact:

A compartment dies when it irreversibly looses its ability to communicate at its highest level of compartmental organization

It is only the *highest* level of compartmentalization that matters: what happens with the lower levels, if present, is irrelevant. In a population, the highest level of organization is reached when individual members communicate. In the vertebrate the central nervous system makes communication possible at the level of the organism: this coordination system is irreversibly destroyed by decapitation. In the eukaryotic cell the highest level of communication is at the level of the plasma membrane; and in the prokaryote or organelle of prokaryotic origin, communication occurs with the outside world across the limiting membrane. The loss has to be irreversible in order to exclude situations of regeneration of the central nervous system like in some flatworms or reversible coma in humans.
First deduction:

The definition of 'Death':

'Death' is the irreversible loss by a given compartment/system of its ability to communicate at its highest level of compartmental organization
This can be symbolically represented as:

$$D(S,t) = -\ C_j$$

where D = death, S = compartment or system, t = time of death, C_j = communication at the highest level (j) of compartmental organization of the system under consideration.

If 'Death' is the irreversible loss of the ability of a given compartment to communicate at its highest level of compartmental organization, it follows that a given compartment starts to live from the moment it acquires the ability to communicate at its highest level of compartmental organization. Therefore we obtain;
Second deduction:

The definition of 'Life':

A compartment is alive when it has the ability to communicate as a whole with its 'environment'. Thus, the very essence of what we call 'life' (as an activity) concerns communication.

Fourth question:

If the very essence of 'life' is the ability to communicate, what then with the other typical properties of the living state as they are mentioned in textbooks on general biology? Are these not essential?

These properties are the five traits that counter disorganization (order, adaptation, metabolism, movement, responsiveness), the three that perpetuate a population despite death of the individuals (development), reproduction, inheritable traits of information or genes, and one important for generating diversity (evolution). To facilitate the understanding of the answer let us consider first an analogy with music. To produce music, one needs an instrument (flute, violin etc.) and one has to play. There are rules that govern the manufacturing of the instrument and there are other rules to produce music once the instrument has been constructed. The majority of the above listed *traits are essential for the construction of the compartment as an instrument for communication*: they are necessary but insufficient prerequisites to make communication possible at the highest level of compartmental organization. Communication requires additional rules as will be explained next.

PRINCIPLES OF COMMUNICATION AND DEDUCTION OF A DEFINITION OF 'LIFE' (AS AN ACTIVITY)

By far the majority of all people who use in daily language the terms 'information' and 'communication' do not realize how difficult it is to adequately define such terms and how complex the machinery required to make communication possible. Before giving an explanation of the essence of communication, let us try to formulate an answer to the following questions.

Imagine first the simple situation: 'A teacher asks a pupil to raise his right arm.' Now consider the questions: If the pupil does this instantly, is this then a situation known in physics as the law of action-reaction, where the energy contained in the command 'Raise your right arm!' equals the energy needed to make the arm conquer the force of gravity? And what if the teacher asks the whole class to raise their arm or asks to do this the next day at 9 o'clock sharp? The insight becomes even more evident if we compare the above described situations with the one in which the teacher would take the hand of the pupil and lift the pupil's arm himself. In both situations the end result is identical: the right arm has conquered the force of gravity.

Information, Communication and Communication Systems

According to specialists in communication theory, already over 150 definitions of communication have been published but there is not much agreement about which one is the best and acceptable in all situations. A general definition is: 'Communication is transfer of information. Such transfer requires a functioning communication system'. 'Information' always deals with a reduction in the degree of uncertainty in a given situation. There are mathematical ways to describe this. A definition of information, suggested by L. Van Poecke, reads: 'Information is any change within a communication system which affects any component of that system'. A communication system functions as follows (Fig. 2).

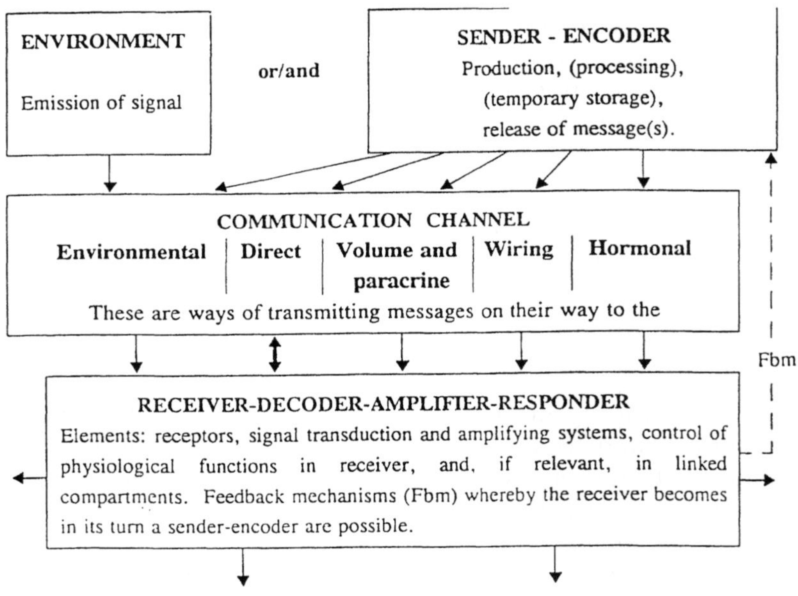

Figure 2. Schematic representation of the different elements of a communication system (modified after [11]).

Either the environment or a sender-encoder delivers some sort of (encoded) message, which is transmitted through a communication channel. There are several types of communication channels: cytoplasmic bridges, nerve axons, blood etc. Next, the message is perceived by a competent receiver-decoder-amplifier-responder. Now comes the essence, *the purpose of communication*. In the receiver (and perhaps as well in compartments interlinked with this receiver through some communication channel), the message, upon having been decoded, *controls the use of a specific form(s) and quantity of prestored energy at a specific time(s).* This energy can then be used for doing some kind of 'work' as this is defined in physics, to engage in other acts of communication (e.g., feedback mechanisms) and/or to cope with entropy (i.e., to prevent the system's break-down).

Let us illustrate this by the example of a mouse trap. If no energy is invested to load the spring in the trap, no mouse will ever be caught. The loading of the spring means prestoring energy that can be used to do work. When the mouse touches the trigger in the trap, the prestored energy is liberated and the trap closes. Touching the trigger is the delivery of information.

The energy present in the incoming message, if present, does not have to equal the amount of energy that will be liberated. There are forms of messages in which there is no energy at all: in some situations absence of something can be information and elicit a reaction. Some forms of information from the past can liberate energy to do work either now or in the future. This means there are elements in communication that allow to anticipate progressing time. This partially explains why 'life' is not fully reducible to physics and chemistry: in information transfer, there can be an immaterial component. Furthermore, it is also interesting to note that in some sense any force in physics can be 'conquered' by communication (e.g., making an atomic bomb explode).

The questions about the physics involved when the teacher asks a pupil or the whole class to raise his/their arm is as follows. The action-reaction law is only operating in the situation in which the teacher takes the hand of the pupil and invests energy that was prestored in the teacher's *own body* to make the pupil's arm conquer gravity. In all other situations it is not the energy present in the command that makes the arm move but energy *prestored in the body of the pupil himself*: the command caused the liberation of this prestored energy.

Communication and Dissipation

Because of their complexity and of the fact communication systems need prestored energy in order to function, such systems are necessarily far from their thermodynamic equilibrium: they are 'dissipative' structures.

At the moment this text is written (1995), relatively few *biologists* in the whole world are familiar with work of the Belgian Nobel price, Ilya Prigogine. In some domains although this matter has been introduced; e.g. Maturana and Varela [12], and Kauffman [13]. However, in my opinion, the nature of 'Life' can only be understood if the principles of his theory about the thermodynamics of far-from-equilibrium systems are superposed on top of the currently known principles and laws of biology [5] [6]. In Nature any system strives to its lowest level of free energy and to maximal disorder: everything will break down in due time and order does not arise out of chaos without investing energy. According to Prigogine, order can be created out of chaos in situations which are far from their thermodynamic equilibrium. Living systems obey all laws of thermodynamics. They manage to create order out of chaos and even increase the degree of order *as long as energy is invested*, thus as long as they are 'open systems'. If the energy supply (feeding) stops, the 'open systems' become 'closed systems' and strive towards maximal entropy (disorder). In the context of 'dissipative systems' , 'bifurcation points' are important. At such imaginary points *choices* have to be made: the possibility to make a choice implies some degree of freedom. It might be speculated that the higher the number of bifurcation points in a given compartment/system, the higher the degree of 'free will' . The number of bifurcation points present in our brain that can be used for communication purposes is just endless. This means the variability which follows hereof is many orders of magnitude higher than that offered by our approximately 100,000 genes and their different combinations. This can be easily understood from the analogy of the endless number of melodies that can be generated with a simple flute with only a small number of fingerholes [7].

Third deduction:

'Living systems cannot be else than dissipative systems with all inherent properties'.

Because 'life' implies communication and because communication requires the presence of prestored free energy and of gradients, living systems cannot else be than situations far from their thermodynamic equilibrium in which order can only be maintained by investing energy. A living system acquires thermodynamic equilibrium when the gradients it uses for communication have collapsed, i.e., when it is dead.

Communication and Chemistry

In cellular communication a specific chemical language is used, communication in living systems supposes biochemical cascade reactions.

The biochemical complexity of 'living' communication systems is very large and cannot possibly be explained in detail in this short chapter. Following example from endocrinology can illustrate this. Oxytocin is a hormone released by the pituitary gland (= sender), a small appendix of the brain, into the blood (= communication channel). In the uterus (= competent receiver) the hormone causes muscular contractions (e.g., those during child birth). The first step in the process is the binding of the hormone, which in itself does not deliver energy, to a specific receptor molecule present in the plasma membrane of muscle cells in the uterus. This binding sets off a cascade of biochemical reactions: the permeability of this membrane for certain inorganic ions (Na^+, Ca^{2+}) increases (specific ion channels will open), this results in a depolarization of the plasma membrane, there is a calcium ion 'explosion' in the cytoplasm, part of cytoskeleton will contract, protein synthesis can be influenced etc. Such 'intermediary processes' are analogous to the different steps in a chain reaction: I call them 'acts of communication'. Not only the qualitative complexity of such acts of communication is large, there can also be quantitative differences. Biological compartments/systems manage to *integrate* ('add up') all their acts of communication. What is called 'life' in biology is in fact a communication activity performed by a given compartment.

Fourth deduction:

A definition of 'Life activity':

'Life' of (dissipative) Compartment/System S at moment t is the total sum of all (coordinated) acts of communication performed by compartment S from its lowest level of compartmental organization (= 1) to its highest one (= j) at moment t.

For a population (level of compartmentalization 13 in Fig. 1), Life (L) can be symbolically represented as:

$$L(\text{population}, t) = \sum_{\text{cell organelle level}}^{\text{population level}} \text{Communication acts}(\text{population}, t)$$

For an animal or plant (level of compartmentalization 6/7):

$$L(\text{organism}, t) = \sum_{\text{cell organelle level}}^{\text{organismal level}} \text{Communication acts}(\text{organism}, t)$$

For an eukaryotic cell (level of compartmentalization 2):

$$L(\text{eukaryotic cell}, t) = \sum_{\text{cell organelle level}}^{\text{eukaryotic cell level}} \text{Com. acts}(\text{eukaryotic cell}, t)$$

In prokaryotes and in cell organelles limited by a membrane, there is only one level of compartmentalization, namely the external plasma membrane (level of compartmentalization 1). Disruption of this membrane finishes life.

$$L(\text{prokaryote}, t) = \sum_{\text{plasma membrane level}} \text{Communication acts}(\text{prokaryote}, t)$$

The same type of reasoning can be applied to any level of compartmentalization (Fig. 1). The general symbolic expression for 'Life' (as an activity) becomes:

$$L(S,t) = \sum_{1}^{j} C(S,t)$$

where L = Life activity, C = communication act, S = system or compartment, j = the highest level of compartmental organization, 1 to j = a given level of compartmentalization, t = the moment at which the acts of communication are considered. Conditions are $\sum^j C(S,t) > 0$ and, when adding up the acts of communication at successive levels, the same act is added only once.

I am aware of the fact that 'time' is difficult to define and may form a major problem in symbolical expressions of the form used here. In my opinion, time is invariably a property of an energy converting system rather than of space (the fourth dimension in the space-time concept). I think that time can be defined as the measure for the inertia characteristic for the conversion, in a given system, of given amounts of one or more forms of energy other energy forms. This means that there are as many different 'times' as there are different energy converting systems; thus endlessly many.

Earlier in this chapter 'death' (D) was defined as the irreversible loss of the ability to communicate at the highest level of compartmental organization and symbolically represented as $D(S) = -C_j$.
Another possibility is:

$$D(S,t) = \sum_{1}^{j} C(S,t) - \sum^{j} C(S,t)$$

More detailed definitions of 'Life activity', but in the same line as the one given here, are possible [7].

Summary of the General Properties of 'Life'

My definition of 'Life' implies:

1. Life is always a property of a *compartment.*
2. The very nature of what we call 'life' is inseparably connected to *communicational activity.* In communication one has to *make a clear distinction between the compartment* used for communication (analogous to a music *instrument) and communication itself* (analogous to the music itself).
3. Life and communication are not possible without prestored energy and without one or more previously formed *gradients.* By definition, a gradient is not a state of stable equilibrium. Exactly because they depend upon gradients, living systems cannot be else than far-from-equilibrium situations, in which order can be created and maintained only by investing energy. Thermodynamically, living systems have the essential properties of '*dissipative systems*'. Irreversible collapse of essential gradients and exhaustion of prestored energy mean the end of life. Viruses are not alive because they do not build up an ionic gradient over a plasma membrane.
4. A compartment can only communicate with the outside world or with other compartments if its *boundary is not perfectly tight*, or, in other words, there must be '*holes*' in this boundary. A cell with a limiting 'membrane' made of glass or an inert metal could not be alive because there would be no holes in such a membrane to let the messages and/or responses pass. Communication at the level of the plasma membrane requires a moderately 'leaky' membrane to make transmembrane ion fluxes possible. Thus possibly, some 3.7 billion years ago, 'life' started as an *imperfection* in a membrane in combination with a chemical gradient over this 'leaky' membrane. The important but usually overlooked role of the cytoskeleton and the distinction between the 'prebiotic' and 'biotic' era's have already been described elsewhere [8].
5. At the level of the plasma membrane, which is the site where the majority of acts of communication are made, communication- and thus 'life' as well- can be considered as the visualization of the incessant *balancing* of the 'living state' *between order* (at the cytoplasmic side) *and 'chaos'* (the environmental side).
6. Life has both *quantitative and qualitative parameters* (types and steepness of gradients).
7. Life started to change and to evolve from its very beginning, and will continue to do so as long as there is life on earth (Evolution): life is never constant. This implies 'life' has a different meaning not

only for each individual compartment, but also at each moment of its existence.

8. Since in many fast acts of communication ionic/electrical processes are involved, *life is to a large extent an electrical phenomenon* [9] [10].
9. In systems consisting of more than one level of compartmentalization, there are acts of communication which coordinate the communication between the successive compartmental levels. It is this coordination which makes the realization of the highest level of compartmentalization functional: this means '*the whole is more than the sum of its constituting parts*', a crucial fact which is often overlooked in the philosophy of reductionistic experimentation.
10. In order to allow a system to transfer and integrate incoming messages it has to contain the 'information' (or programme) needed to do so. Therefore, living compartments need to build up *some kind of memory* (short-term and/or long-term storage of information).
11. Under proper conditions communication, thus life, can master any force known in physics.
12. Life is above the chemistry and physics required to form the compartments, the instruments for communication. The *immaterial and indeterministic* aspects of 'life' *primarily result from the laws operating in dissipative systems and communication, less from those of biochemistry and genetics.*

IS A COMPUTER ALIVE? 'ARTIFICIAL' OR 'MAN-MADE' LIFE

Some man-made mechanical and/or electronic devices (or robots) have become so sophisticated they start to look like being alive: e.g., imagine a man-made computer in which the electrical energy comes from a solar cell. A system like this can have all the necessary elements of a communication system as outlined above (stockpiled energy, a built-in programme to act as a receiver-decoder-amplifier-responder and the possibility to use energy for carrying out specific types of work). When a signal is transmitted via a keyboard, the system will react. In principle, such a system is not basically different from the one in which a hormone binds to its receptor (= the equivalent of typing the right code on the keyboard) which initiates signal transduction to mobilize previously stockpiled energy for doing work. Intuitively everybody will reject the idea the computer is alive, although we sometimes use a similar vocabulary: e.g., we say the computer is 'dead' like we say the radio is 'dead', meaning that such communication systems failed. If one analyses the system, one has to conclude we do not have good arguments to firmly say a computer is not alive (for criteria, see [7]). In computers, we are dealing with another form of life, man-made, and not based on carbon-chemistry and electricity carried by inorganic ions like in true biological systems, but on silicium-metal chemistry and on electricity carried by (the much faster) electrons. It is possible to adapt the symbolic expression for life given before in such a way it becomes possible to make the distinction 'normal biological' life and 'artificial' life:

$$L(S_{(TC,TE)},t) = \sum_{1}^{j} C(S_{(TC,TE)},t)$$

in which TC = Type of Chemistry, TE = Type of Electricity of Energy in general, the other symbols being the same as in the first symbolic expression of life.

EVOLUTION: 'LIFE AS A DOUBLE CONTINUUM'

From my way of defining life it follows that the neo-Darwinistic approach of evolution can only partially describe the mechanisms that govern evolution because this theory considers *life as a single continuum*, namely the continuum of the compartments (= instruments) needed to make communication possible. In neo-Darwinism mutations, changes in gene frequencies in populations and selection are the essential causal elements in evolution. However, *life is not a single but a double continuum*: first, there is of course the continuity of the compartments, but second, there is also the continuity of communication and information themselves: certain types of information can be transferred from one individual to another or/and from generation to generation through teaching and learning, this being partially independent from genes. What is known as 'cultural evolution' is in fact nothing else than communicational evolution and such evolution is not at all restricted to the *Homo Sapiens*. In this double continuum theory of Evolution [7] elements of communication have evidently an important role: *in this theory the 'communication force' drives evolution and selection is rather a result than a cause.* The symbolic representation of Evolution (Ev) in its most simple form is:

$$Ev(S, t_2 - t_1) = \Delta L(S, t_2 - t_1) = \Delta \sum C(S, t_2 - t_1)$$

$(t_2 - t_1)$ representing a span in the geological time-scale.

FINAL CONCLUSIONS

In my opinion biology has reached such a degree of maturity that it is ready to undergo as revolutionary a change in thinking as physics and chemistry experienced decades ago when the non-predictable quantum physics-theory was introduced to complement the deterministic Newtonian physics. Classical biology as it is taught today is in fact as deterministic as Newtonian physics [4]. Superposition of the non-predictability inherent to communication and dissipative systems on top of contemporary biology finally allows to define 'Life' and make fit together in a most logical way all the different parts of the 'biological puzzle'. For this novel approach in biology [12] [13], De Loof proposed the name 'communication' biology [14]. Quite some established facts and concepts will need rethinking and reformulating. This type of biology allows to link the biology of 'compartments as instruments for communication' (= what contemporary biology is all about) to psychology and sociology, both being sciences which primarily deal with communication itself. It is not at all uncommon, as I know from my teaching experience, students in psychology and sociology consider contemporary biology as being of little value for a better understanding of their disciplines. In dissipative biology, because it focuses on communication, there is ample room for emotions, feelings, beauty, freedom in acting, and higher values, which is not the case in classical biology.

Arnold De Loof
Department of Biology of the K.U. Leuven,
Zoological Institute, Naamsestraat, 59,
3000 Leuven, Belgium
e-mail: arnold.deloof@bio.kuleuven.ac.be

REFERENCES

[1] **Schejter, A., Agassi, J.**, "On the definition of life", *Journal of the General Philosophy of Science*, **25**, 1944, p. 97.
[2] **Schrödinger, E.**, *What is Life?* Cambridge University Press, Cambridge, 1944.
[3] **De Duve, C.**, *Blueprint for a cell. The Nature and Origin of Life*, Neil Patterson Publishers, Burlington, USA, 1991.
[4] **Cramer, F.**, *Chaos and Order. The complex structure of living systems*, VCH Verlaggesellschaft mbH, Weinheim, New York, Basel, Cambridge, Tokyo, 1993.

[5] **Prigogine, I., Nicolis, G.**, "Biological order, structure and instabilities", *Quarterly Review of Biophysics*, **4**, 1971, p. 107.

[6] **Prigogine, I.**, *From Being to Becoming. Time and Complexity in the Physical Sciences*, W.H. Freeman and Company, San Francisco, 1980.

[7] **De Loof A., Van den Broeck, J.**, "Communication: the key to defining "Life', "Death" and the force driving evolution. "Organic chemistry-based-" versus "Artificial life"", *Belgian Journal of Zoology*, **125**, 1995, p. 5.

[8] **De Loof, A.**, "Schrödinger 50 years ago: "What is Life?" "The ability to communicate", a plausible reply?", *International Journal of Biochemistry*, **25**, 1993, p. 1715.

[9] **De Loof, A.**, "The electrical dimension of cells: the cell as a miniature electrophoresis chamber", *International Review of Cytology*, **104**, 1986, p. 251.

[10] **Harold, F.M.**, *The vital force: a study of bioenergetics*, W.H. Freeman and Company, New York, 1986.

[11] **Geraerts, W.P.M., Smit A.B., Li K.W.**, "Constraints and innovations in the molecular evolution and neuronal signaling: implications for behavior", In: **Greenspan, R.J. and Kyriacou, C.P.** (eds.), *Flexibility and constraints in behavioral systems*, John Wiley and Sons Ltd, New York, 1994, p. 209.

[12] **Maturana, H.R., Varela, F.J.** *Autopoiesis and Cognition. The Realization of the Living*, Reidel, Dordrecht, 1980.

[13] **Kauffman, S.**, *The Origins of Order. Self-organization and selection in evolution*, Oxford University Press, Oxford, 1993.

[14] **De Loof, A.**, *Wat is leven? De onstoffelijke dimensie*, Garant Publishers, Leuven, Apeldoorn, 1996.

SERGE GUTWIRTH

SCIENCE AND ENVIRONMENTAL LAW. THE PERSPECTIVES OF TRUTH-PLURALISM

INTRODUCTION

Today, no one seriously doubts that the sciences are crucial to ecological problems. They are involved in the examination and policies of ecological issues, and occupy a central position in the relationships between culture and nature. As a consequence of that position, scientific concepts emerge in the legal field and call for a sound reflection about the relations between science and law, fact and law, scientific truth and justice [19]. Such a reflection appears particularly urgent when it comes to questions on environmental law [20, 21, 23, 24, 44].

This contribution takes a theoretical and prospective approach. It is geared towards the construction of a transdisciplinary conceptual framework, which might trigger a reconsideration of the relations between the sciences and (environmental) law in the light of conceptual shifts in science or meta-science. The contribution does not provide an exhaustive description of the multiple interactions between science and the practice of positive (environmental) law. It rather presents a meta-narrative about those interactions.

The analysis proceeds in three stages. The first stage deals with the sciences in general. It describes a number of changes which their modern conception underwent under the impact of 'truth-pluralism'. The analysis then moves on to tackle the impact which these changes were seen to have on the overall relation between science and law. The consequences of these changes and their impacts on the relations of science and environmental law are explored in the last stage of the analysis. Three paths of thought are suggested, that are expected to inform a fundamental reflection of the relations between the human and the non-human or between culture and nature.

© 1999 *Kluwer Academic Publishers. Printed in The Netherlands.*

CONSIDERING THE SCIENCES: FROM A MODERN CONCEPTION TO TRUTH-PLURALISM

The Modern Conception of Science

Three assumptions are believed to support the modern conception of science.

First, the principle of order and rationality must be invoked: modern science presumes a mechanical world, that is ordered and 'legal'; a world that is seen to function according to the principles of reason. Following Galilei and Newton, modern science has postulated the existence of general laws, that are invariable (eternal) and universal. By principle, it assumes a world that behaves in a mechanical way, as if it were an automaton [49].

Second, modern science can also be seen to incorporate the Cartesian credo, according to which man—and especially the scientist—has a duty to *master and own* nature (*maîtrise et possession*). Man's power over the non-human is not problematic to that concept, it is *given*. Since man is the only being blessed with rationality, he is the only being capable—and therefore entitled—to discover and master the rationality of nature. Rationality is thereby projected onto nature, so as to give the scientist, as Isabelle Stengers writes, *un droit de la raison* that underwrites his power over the phenomena of nature [55, 56]. In sum, Western modern science—synthesizing rationalism and empiricism—assumes the existence of a mechanical world, which is ordered, which can be known in an objective way and which should be mastered and controlled by humankind.

The third assumption of modern science is recognized in its practice of inferring a rigid separation between subject and object, and between culture and nature. That conceptual dividing line supports the claim of modern scientific practice that it holds a unique position vis-à-vis the objects: only by way of that relation can these objects be discovered and revealed *as they are*. This is the principle of *adequatio rei et intellectus*, according to which the scientist can define himself to be the neutral, uncolored and passive spokesman of the object. By that title, naturally, he also posits himself out of society. Here then is the reason why science can pretend to be objective and true, and can portray itself as independent of political, social, cultural or moral values. That portrait is fed into the scientific enterprise, thereby encouraging its belief that it can truly escape from the political mediations that characterize all intersubjective relationships. Moreover, scientists take themselves to be the *only* mortals that are allowed to make that claim: only science can reveal the truth about the object. The sciences, as conceived in the modern way, therefore proceed on the basis of what the French sociologist of science Bruno Latour has termed *les Grands Partages* (the Grand Divides). One such *Grand Partage* separates subject from object, another separates science from everything-but-science [33].

In the light of these three assumptions, science can claim a vast and arrogant power. First it appropriates a right to what it claims to be the only knowledge 'that is really true', that is to say, the absolute and unsuspicious truth. It then justifies that claim by another one, that is a claim to have an objective relation vis-à-vis the objects and to be the neutral spokesman of rationality. Finally, science dresses itself in the 'absolute difference' by way of which it situates itself entirely beyond the realm of society, politics, the economy, ethics and law. This argumentation legitimates the unlimited power that science claims over the object(s). That power, moreover, is doubled by its power vis-à-vis other humans and human affairs, namely the power to say 'what cannot be doubted or even be discussed', a power to be unsuspected [54]. In other words, the modern conception of science bans every kind of political or legal mediation from its territory.

Science in the Light of Truth-Pluralism

Today, nevertheless, the modern conception of science is put to the test. It is questioned both by its own internal progress, and by meta-scientific activity in anthropological, social and historical studies of science. These findings are summarized in the next section, and can be seen to proceed from a reorganization of several themes that outline a new approach to science, and that I will henceforth indicate with the term 'truth-pluralism'.

Scientific practice has collided with the limits of its own modern assumptions, and thereby constitutes the first pillar of truth-pluralism. Science has demonstrated 'from inside out' that order and rationality can no longer be considered as the building blocks of the universe. Order appears to be rather exceptional, chaos being the more natural or regular state of affairs. Serendipity, indeterminism, uncertainty, relativity, self-organization, and still other concepts have entered the conceptual domain of science as well. These have all added to the conclusion that, contrary to what the modern conception of science has claimed, the real is not rational.

Time has been discovered to be irreversible. This second scientific discovery of great conceptual importance, in its turn, was seen to imply that matter has a history, and that each set of changes in that matter necessarily redefines the conditions for the next set and change. Each event redistributes the given.

Furthermore, Heisenberg's demonstration must be kept in mind: the observer and his techniques of observation disturb the object of observation. The observer acts as well. More generally, science has realized and recognized that man is an integral part of the world. He is marked by that world, and he marks that world. He is at least partly determined by his environment, that he determines in his turn. Subject and object are no longer conceived as separated by a sterile wall. Rather on the contrary, humans and nature are taken to be interwoven, human history and natural history are one, and subject and object are both caught up in one inevitable game of interferences and interpenetration.

All of these insights erode the claims of science towards absolute truth, one sole truth that is timeless and that resides beyond society. Indeed, since subject and object interact, each establishment of a scientific truth finds itself linked up with the actions, projects, events and powers that it has enabled. The modern conception of science finds itself replaced by a plurality of truths, where each truth is just as 'hard' and objective as it may turn out to be, and is just as general as it may seem, and holds for as long as it turns out to hold. Truth is thereby marked, once again, by its own conditions of existence. That awareness affects both the power that subjects hold over objects, and the power that science holds in the social arena [19, 24, 44, 49, 54].

The critiques that the scientific enterprise has made on the modern conception of science have been consolidated by 'external' critiques. The latter have been formulated by the anthropology and/or sociology of the sciences, which constitute the second pillar of truth-pluralism. This pillar refers to the works of researchers who have been portrayed as 'the galaxy of disciples of Michel Serres' by François Dosse [12]. Anthropology of science, as practiced by *inter alia* Bruno Latour [9, 30, 31, 32, 33, 34, 39, 56] examines what scientists actually 'do' rather than what they pretend to be doing or what they are supposed to do. These studies have actually provided strikingly little evidence of the image that science proposed in the modern conception. Rather on the other hand, these studies of scientific practice have illustrated that facts are actually 'made' or 'constru(ct)ed'. *Les faits sont faits*, as it has been put by Bachelard and Latour. In the gaze of the anthropologist, scientists are revealed to be heavily involved in strategic activities. These are geared towards ensuring that their particular viewpoints and arguments become prominent and accepted in the established scientific networks. These networks, in their turn, appear connected with non-scientific networks that encompass politicians, the military establishment, industry and other sponsors.

In order to be proven right scientists incessantly combine and mobilize arguments, ideas, concepts, things, techniques, machines, institutions, and other allies. Scientists are seen to produce scientific facts from these elements and strategies, in the course of a typical game that involves construction of references (*construction de références* [36]) and reciprocal testing and putting to proof (*mise à l'épreuve réciproque* [56]). Scientific facts are the products of a human activity, and yet they are taken to correspond with the untouched reality of the objects. What(ever) started out as an argument, project, program or controversy, finds itself transformed into a fact or a scientific truth at the end of the day. 'La cause devient une chose' as has been observed by the French scholar Michel Serres [1]. As soon as the controversy is closed, the facts and scientific concepts can be introduced into the social arena as if these had always existed, as if these had no history, as if these only had to be 'discovered'. Objectivity, neutrality, universality and the truth of scientific claims are, therefore, the product of an utterly human activity, that is nervous, polemical, and uncertain. Modern science has always purified its scientific truths from the imprints of the projects that preceded the ascension of these truths to the status of science.

All of these observations provide a rather surprising image of science. On the one hand, science finds itself founded by two modern Grand Divides. Quite on the other hand, one now realizes that mixed forms proliferate in the real world. That real world is seen to consist of (the results of) mediations between objects and subjects, between science and non-science. This conclusion is unnerving, since it reveals the modern conception of science to have teamed up with an activity that science itself holds impossible, that is the production of what Latour has coined 'hybrids'. Even worse, the modern conception of science occludes the production of hybrids from the eyes of the observers and thereby withdraws that activity from the democratic processes of a legal and political kind. From this perspective, it becomes clear why Western modernity never provided for a place where science is called to account for what it is doing. The deceptive image that modernity created about the sciences must be undone, without however undoing scientific practice itself. To that end, we must accept 'never to have been modern', as Bruno Latour has already shown. In his view the hybrids take up too much space, they can even be said to occupy the entire territory. The task, then, is to turn the 'undivided' into the first and founding principle: in science, as elsewhere, the subject is bound to collide with the object, and science itself necessarily interacts with society.

Consequently, science must be reconceived as an open-ended and collective activity, that is performed in socio-technical networks. One should not hesitate to acknowledge that these networks have political and legal dimensions [56a]. Since scientific truths and facts are the outcome of projects, their representation must be considered quite like the political representation of men, that is as a conflictual and contestable form of political representation. Actually, scientists, just as political representatives, may be more likely to represent themselves than the objects whose objective spokesman they pretend to be.

All of these observations provide reasons to believe that the regime, whereby the sciences are held to represent objects by the principle of *adequatio rei et intellectus*, ought to be converted into a regime that is inspired by political representation and its typical features. That last regime *explicitly* provides for processes of delegation, for democratic discussion and for a transparent legal framework [2]. Were these principles of political democracy to be applied to both forms of representation, scientists would be seen to represent hybrids in the same vein as humans represent other humans. On these grounds, Bruno Latour has introduced the concept of the *Parlement des choses* ('Parliament of Objects') [33, 36] that could render account more openly of what already exists, albeit in a clandestine way, in scientific networks today, that is controversy, negotiation and mediation.

The concept of *Parlement des choses* is unambiguous; it expresses that scientific research has a political dimension. Nevertheless, scientific representation of the objects remains in the hands of scientists, even if the latter are no longer taken to be experts in the narrow meaning of that word: they have rather become political delegates, who can effectively be made to account for their standpoints. They now enter on stage without the aureole, in the midst of doubt, they can no longer appeal to be beyond doubt. Their absolute truth claims have shrunken to reasonable proportions, and their debates will henceforth pertain to projects, relations of power and actions, rather than to 'facts' and 'truth'. In such a constellation scientists have to confront their truth claims with other truths that live in society. The *Parlement des choses* assembles scientists, policy makers, politicians and administrators to jointly take up responsibilities that spring from a threefold demand of quality: democratic, scientific and practicable [36]. The *Parlement des choses* calls for "une dialogique généralisée impliquant tous les porte-parole concernés" [12a].

This design seeks to dismantle the system whereby irreversible political decisions are made on the basis of unquestionable and unquestioned scientific data. In its stead, a new framework of management and political navigation is envisioned, that is likely to reduce and slow down the production of hybrids. It must provide for delays, moments of reflection by biasing procedures that trigger or even impose controversy and concertation, and thereby ensure a sound balancing of interests, values and truths. This practice of political and legal temporizing is conditional to the introduction of democratic schemes into scientific networks and into our attitude towards nature. Moreover, it is legitimate because it imposes caution, measures of precaution and therefore responsibility. The *Parlement des choses* refuses accomplished facts and absolute political decisions.

None of these observations and suggestions prevent that we recognize and even reaffirm the singularity of the sciences and the special status of scientific research. The aim of the argument is not to reduce science to 'nothing but politics', nor to fall into a kind of relativism that turns each scientific claim into 'a mere opinion' [3]. Such attitudes would fail to explain the singularity of the enterprise that constructs reality and that we tend to refer to as science. On the contrary, it is easy to indicate what makes science special as compared to other human activities. Scientific research is essentially characterized by a constant putting to test of its conclusions within the confines of its own networks. As Stengers writes: "Dès lors qu'il s'agit de science, tous les énoncés humains *doivent* cesser de se valoir, et la mise à l'épreuve qui *doit* créer une différence entre eux implique la création d'une référence qu'ils désignent et qui *doit* être capable de faire la différence entre science et fiction" [4]. Anthropology of science describes research as a genuinely specific and distinct activity. In its terms, the representation of the non-human passes through the construction of networks, that must constantly consecrate and confirm the authority of the representatives who are constantly being put to the test by all other actors in their network [36].

Why then coin the term 'truth-pluralism' to describe this new approach to science? First, 'pluralism' suggests that the new approach does no longer allow for one unique truth about the facts, nor for a scientific power that is above all doubt. It only allows for plural truths that are relative, or better even: it only allows for plural truth claims. To paraphrase Claude Lefort with René Foqué, the space for 'the truth' is necessarily left empty [19a]. A second argument to use the term 'pluralism' is that it does justice to the multiple controversies, discussions, and falsification efforts that characterize 'science in action'. There is a political dimension which is constitutive to scientific work. Since we can discern projects and choice, power and resistance, winners and losers in each junction of the network, it must, therefore, be subjected to the rules of democracy and the constitutional state (rule of law). Pluralism is inferred in two ways: first, it implies the recognition of the rights of the losers, who are no longer excluded of the debate but find themselves represented as a minority group, and second it underlines that each political and legal project must be pluralist if it is to be democratic [57, 11, 22, 28].

The concept of 'truth' remains in truth-pluralism, since this seems to be an adequate strategy to do justice to the singularity of science and research. The reciprocal putting to the proof, the incessant falsification practices, the mobilization of arguments that are 'harder' than other arguments, the construction of experimental set-ups, and the construction of references all characterize science as an activity that generates a specific type of knowledge that tends to be exact, hard, stabile and relevant. The use of the word 'truth' then, is a tribute to that particular project. Truth-pluralism thereby points at a practice that ponders, evaluates and discusses the multiple truths and interests that are at stake. That concept implies that it necessarily takes time to reflect, evaluate and judge the hybrids that science has given out. That production must be slowed down by channeling it through processes of evaluation and consideration, that are henceforth transparent, visible and open. Whereas science once possessed the truth by law, could proceed at maximum speed, did not look back, and fled over each obstacle it encountered along the road, it is now anchored into society where discussion, controversy and politics call for responsibility, prudence and caution.

RECONSIDERING THE LINKS BETWEEN SCIENCE AND LAW

The evolution of a modern conception of science towards truth-pluralism necessarily affects the relations between science and law. Two moments must be distinguished in that respect. The relation of 'modern' science and law is discussed first. The advent of truth-pluralism in law is analyzed later, in a prospective mode.

First Moment: Modern Science and Law

This first moment can be symbolized in terms of two icons. The first is Galilei. The legend portrays him at the time of his condemnation, while making his heroical statement: "eppur si muove!". The message is clear. Whatever the tribunal may think, scientific facts take priority, that is, in this particular case: the earth turns round. As a second icon we may invoke the principle of 'free examination' that both the *Université Libre de Bruxelles* and the *Vrije Universiteit Brussel* have adopted as their motto. It demands that "la pensée scientifique ne se soumette jamais, ni à un parti, ni à une passion, ni à un intérêt, ni à une idée préconçue, ni à quoi que ce soit, si ce n'est aux faits eux-mêmes" [50a].

Both cases clearly posit the truth of science beyond the realm of social processes and their political and legal dimensions. They infer a world that escapes from human judgment and that can only be known by the voice of science. That inference is quite something! Not only does it translate the modern pretension of the exteriority of the sciences, it also ascribes to the idea that the legal realm must be submitted to science, that Justice should be subjected to Truth. This position consecrates a purely instrumentalist view of law in which its role is reduced to the mere enforcement of scientific truth. Modernity privileges science to such extent that even the law must submit to it and must turn to science to fill in its contents. Such relation between the sciences and law has—regretfully perhaps—been more than symbolical. It is exemplified in several branches of positive law, where scientific truth has been seen to mark the contents of law to a striking extent [19].

This subordination is problematic in two ways. As we have already seen, the implicit modern conception of science is never evident. The absolute, extra-political and indisputable role that it has awarded to science has been revealed to be quite unfounded by truth-pluralism. But meanwhile, it is painful to see what a poor idea of law is expressed through this relationship of modern science and law: a law that has no autonomy, no content of its own and which is reduced to a mere instrument of the sciences, whereas the essence of law is its role to be a mediator between different powers in order to curtail every kind of absolutist excess [16, 17, 18, 19, 20, 22, 25, 26, 27, 28]. The legal system, in the true meaning of that word, must therefore provide an autonomous 'relational' framework that befits arbitration and mediation, and can weigh out different interests (cf. Foqué & 't Hart [16, 25]). Such framework, quite obviously, must leave its concepts open, which implies that it must never be faced with a discourse of one sole truth. And yet, the discourse of modern science claims a truth that is apolitical and ahistorical, that resides beyond doubt and above the law. Under these circumstances, the system of law finds itself transformed into a weapon of the actors and truths which have already triumphed within the scientific network. Instrumentalized, in this sense, it loses its quality of law. Such law merely (re)legitimates the discourse of science. If, however, the law were to be understood as a set of open concepts, it qualifies as the outstanding place to mediate between narrative and dogmatic types of knowledge, between the factual and the abstract, between the local and the global, between the multiple and the unique. In that position it would effectively, which is its primary function anyway, come to mediate between the power on one hand, and the resistance on the other. There is no reason to skip this power-balancing with respect to powers that the sciences legitimate (cf. the complete works of Michel Foucault).

Second Moment: Truth-Pluralism and Law

The problems that have been outlined above call forth the second moment, that is to say the rather obvious overture of truth-pluralism and the relational concept of law. The overture is obvious since law and truth-pluralism have a common viewpoint on truth claims. Neither the one nor the other can accept a narrative that is absolutely true, that stands out above the law (and judgment), or that claims to represent the only conceivable law (or judgment). Today, both law and science must acknowledge that all truth is rooted in a social network, in projects, strategies, aspirations, conflicts and therefore: that each truth claim is ensconced in a network of power relations.

The 'fact' being constructed, it translates a project with its values and implicit norms. This last observation explains the overture of law and science. It is about time to depart from the polarized image of a normative law ('ought') that is confronted with a descriptive science ('is'). The law, and case law even more so, must refer to the facts and reality, even if these can no longer be described in terms of absolute truth [14]. Law can only exist if there is some space left to mediation, but there is no meaningful practice of law without truth-seeking. Underdetermined (or 'contrefactices') concepts must therefore support a relational type of law. Conversely, truth-pluralism introduces the political and legal moment—norms, choices, projects, values—into the heart of scientific activity, even if that activity only seeks to be descriptive. Truth-pluralism entails and enforces a relational approach to law, because it demonstrates the impossibility of absolute truth claims that could be appealed to legitimate an 'imperativist' view of law. Meanwhile, the relational conception of law can but favor the acceptance of truth-pluralism, because it cannot, in any way, signify the end of science and research. Legal concepts and scientific concepts may even be said to fit together well, since both are contextual, open, plural and tied to their conditions of existence.

The two cases that were invoked in the beginning of this section, and the analyses of positive law that demonstrate their pertinence, have shown the relation of law and modern science to be aversive to truth-pluralism and to enforce the erosion and instrumentalization of law. Legal practice is effectively eroded when 'modern' truth claims are imposed upon law, and rob the latter of its essential aspect: If the law is to enable social plurality while ensuring, at the same time, the perseverance of the respectable 'order' of liberties, diversity, differences, then that law must always remain the mediator. Truth-pluralism in the sciences can be understood to bring rather good news for democracy: The autonomous action of relational law might as well be moulded in the same spirit of innovation. It is no longer possible to conceive of a truth (that, by the way, is necessarily linked up with power relations) that can escape the legal temporizing and mediation process, neither in the human sciences nor in the natural sciences. No single truth claim about the essence of man and his society can henceforth occupy the entire space: alterity, difference and plurality remain the obligatory perspective and fundamental objective. Neither can a single truth describe the environment, nature or the objects in an absolute way. This implies that it is no longer inconceivable to think the political-legal moment in terms of the relations between humans and non-humans [44]. No one doubts that this last perspective can only be met with satisfaction.

SCIENCE AND ENVIRONMENTAL LAW: WHICH KIND OF RELATION? WHICH KIND OF DIALOGUE?

An Encounter that is Privileged ... and Paradoxical?

These conclusions on the relation between science and law in general apply to the relation of science and environmental law as well. Environmental law is even said to be one of the privileged spots where science and law can meet (another good example being the so-called 'computer law'). Quite often, only science can show, identify, and formulate the ecological problems to which environmental law must find the answers. Scientific endeavors are also found at the basis of environmental legislation and norms: quite often these are, after all, 'socio-technical norms' [8]. The sciences, again, have a fundamental role in the application and practice of environmental law. Finally, the sciences intervene frequently and in a decisive way in juridical procedures and environmental matters.

That impact of the sciences may appear paradoxical. As a matter of fact, scientific progress has created the opportunities for the scale on which the ecological crisis is taking momentum today. Are we, then, to cure evil with more evil? Moreover, if science is called upon to resolve environmental problems, that call runs counter to the ecologist discourse that used to badly accuse science and technology. Pierre Lascoumes acknowledges that issue as the 'revenge' of the old scape-goats, that is to say, the scholars, engineers and experts. This 'return of science' can even be expected to advance an eco-power that they are putting up, assuring, moving and holding in their hands, and by which they seek to guide and determine the political decisions that we are taking about 'the living' [40].

In the light of the road already taken, the paradox can be lifted. The matter certainly looks like a paradox if one envisages a 'modern' and thereby outdated type of science. But wouldn't its outlook alter considerably if the science concerned were the kind of science shaped by truth-pluralism? Ecology and the new approach to science have emerged simultaneously. To a certain extent they have jointly contributed to the changes and evolutions in today's socio-technical and scientific networks. Rather than bringing scientism and modern science back to power, ecology presumably participated in clearing the road to greater openness, democracy, representativity, modesty and, in short, to truth-pluralism in science. If that is the case, the paradox appears less manifest. The come-back of scientists might well be a come-back of a new generation of scientists who are marked by the perspectives of truth-pluralism. From that viewpoint, the 'return of science' turns out to be good news rather than a revenge. It effectively opens up promising perspectives to environmental law and to an even more encompassing reconsideration of the legal relations between men and the non-human.

The shift from a modern conception of science to truth-pluralism may well incite to democratic reappropriation of environmental law and politics. We must not lower our arms when confronted with an *éco-pouvoir* but explore the perspectives and concepts that enable more balanced, open and promising relations between politics, science and law. Since there is no power without resistance (as showed Michel Foucault), since democratic law has the duty to protect that resistance which is therefore legitimate (supra), and since finally, the novel approaches to the sciences open up the networks of knowledge production to the public eye, the clandestine status of these networks may alter in a fundamental sense. As a result, politics and perhaps even democracy may emerge in the scientific networks.

And yet, this might be jumping to conclusions. Unfortunately, little optimism is due with respect to the current legal organization of the relations between men and nature. These relations are not implicated with truth-pluralism, but appear strongly marked by modern science.

The Pessimistic Conclusion: Isomorphism between Environmental Law and the Approach taken by Modern Science

The analysis of the current legal relations between man and nature cannot be restricted to a mere examination of environmental law, because these relations are ruled, in their turn, and before anything else, by ownership and property law. Property is the right to enjoy and to dispose of things in the most absolute and exclusive way. In the French *Déclaration des droits de l'homme et du citoyen* property is defined as an inviolable and sacred right. That approach obviously corresponds with the Cartesian idea of *maîtrise et possession* and with the principle of the *Grand Partage*: man asserts his absolute power over things to which he is considered entirely exterior. Indeed, laws and case-law constructions exist that limit the absolute character of the property regime. These restrictions, however, are always seen to be imposed in function of *human* and social interests. As far as an object proper is concerned, there are no limits. Restrictions to the rights of the owners are never justified by a project to end the damages that all the owners together cause towards everything that can be appropriated. As Michel Serres describes, the law is exclusively concerned with wars between humans (*guerres subjectives*), while it is never turned towards the world into which these wars are fought. There is no mediating law, no arbitration or conciliation procedure. Nothing protects nature against the objective violence (*violence objective*) caused by the hostilities of the united warriors, who are fighting their political, economic, military, technical, scientific wars [52b]. The subject exercises its power and ownership as if these were given to him 'by nature' or by the order of things. Consequently, the entire group of owners exercises a power over the non-human that is out of control, boundless and arbitrary. In this sense, property law and ownership clearly correlate with the modern assumptions of science. The subject dominates the passive object in two ways, epistemological and legal. This double power relation, this isomorphism between modern science and law, has been understood to lie at the origin of the present ecological crisis. It is there, as Michel Serres has shown so clearly in *Le contrat naturel*: "Voilà pourquoi nous vouons nécessairement les choses du monde à la destruction. Maîtrisées, possédées du point de vue épistémologique, mineures dans la consécration prononcée par le droit" [52b]. That same isomorphism of legal concepts and modern science has also been revealed in the system of intellectual property law and know-how protection [13, 19, 46].

Environmental law as such, unfortunately, cannot be concluded to have escaped from the assumptions that are common to the modern conception of science and property law. The expectations that it ever would were probably too high, since these implied environmental law would effectively confront the underlying principles upon which our societies organize the relations between men and objects. Respect for property, for the 'laws of the market' and for the epistemological principle of *maîtrise et possession*, had too much of a head start as compared to the young branch of law that now seeks to limit these principles for the benefits of the environment. In the present legal constellation, therefore, environmental law is grafted on a legal stem (property law) that does not support the realization of environmental aims. That stem, moreover, is rooted in a concrete and given power situation of subject over object, which has traditionally reinforced the pretensions of modern science. All of these facts are translated by the well-known lack of effectiveness of environmental law, by reports of its non-application, by its various practical problems and by its lack of theoretical foundations. Environmental law is hampered by the modern conception of science and by the unilateral relation between man and the objects that surround him. Under these constraints, it is almost bound to be incapable of intervening in the domain that generates and maintains the power of man over the non-human. This explains why environmental law seems condemned to do no more than clear up or readjust the worst effects of the damages that have already been done [10]. Environmental law is in serious trouble to juxtapose itself to the system of property law: it does not innovate that system in a fundamental way, it does not even shake it up. Environmental law has entered the game long after the cards were handed out, after the scientific, legal, and economic power relation of the subject over the object had been established. These insights provide very little reason for optimism because environmental law is likely to function merely as an exception to the principles of property law.

The Optimistic Perspective: a Fundamental Reconsideration of the Law that Guides our Relations towards the Environment, in the Light of a Truth-Pluralism

Must we stay then with that conclusion? No, and on the contrary, since truth-pluralism has been expected to engender a fundamental reconsideration of our legal arrangements vis-à-vis objects, the environment, and the ecological stakes. That new approach can alter the relations between the scientific networks and the political-legal universe on the one hand, and the relations between man and the non-human on the other. These possibilities must be explored in the light of two targets. A first target lies in the establishment of democratic decision-making procedures in ecological issues. The second target calls for optimizing our use of scientific knowledge (and uncertainties) and of diverse interests that are concerned. Environmental law can eventually be reconsidered in the light of that exploration.

Ecology and Truth-Pluralism

The development of truth-pluralism has been portrayed as concurring with that of ecology, the science of ecology and the awareness of ecological problems [41, 42]. The problems and the controversies on ecological matters have called for scientific reflection in several new dimensions. These have urged it to pass on from the local to the global, underlined the risk that is inherent in each innovation, provided a deeper insight in the complexity of the phenomena and processes that are observed, into their ultimate effects in time, into the irreversibility of these processes, and so on. In addition to these new dimensions, some relations have been reversed entirely: the aim is no longer to control nature, but to control the control. According to Serres a new *maîtrise de la maîtrise* should replace the old *maîtrise et possession.* Science must no longer occlude its political and legal aspects, it must help design new forms of representation and decision making processes. In sum, ecological problems simultaneously question science, politics, the economy, morals and law, and thereby shake up the modern construction of the *Grands Partages.* When facing up to the questions of the environment, the domain of law and politics finds itself penetrated by science, and vice versa [36, 37, 38].

New Questions

A new set of questions proceed from an articulation of truth-pluralism and a legal practice that must face the ecological crisis.

How can an environmental law be set up, if we have no more reasons to speak of 'one environment' that is external, passive, mastered and owned by man, and instead must refer to the interactions, networks, and to the constant interweaving of people, things and discourse, that Latour has termed 'hybrids' and that François Ost has called 'projects' [46]? How to establish our relations towards the non-human, if we can no longer infer a nature in an essentialist tradition, a nature that has no history, that stands beyond politics, is unmovable, eternally invariable, but instead must refer to that non-human realm in terms of active processes, that engender certain environments from the relations between innumerable human actors and the non-human world? How to think our relations with nature and objects beyond the *Grands Partages* and the antagonistic concepts of subject and object? And how to do away with the mutually antagonistic ideologies of anthropocentrism and ecocentrism ? How to conceive of a 'mid-space' which truth-pluralism has shown to occupy the 'entire space'. And what words and legal vocabulary must we invent to that end?

Even if we cannot provide full answers to these questions, we can indicate the following paths to direct further reflection.

Paths of Thought

The first path underlines the necessity to consider the relations of humans and nature as political and legal relations, that must be managed in a democratic and open way *and* in real time, in the light of scientific uncertainty as well as with the certainty that human and natural histories are interwoven. The ongoing collective experience must be steered in a democratic way, and a protocol must be considered to that end [38]. In other words: the production of hybrids and projects must be opened up to debate, negotiations and political discussions. That strategy will dismantle the current clandestine functioning of the networks, and eventually define the power relations as what they are. The expert mode and purification practices must be altered, so as to befit a *Parlement des choses* that could realize the passage of modern science toward a mode of truth-pluralism.

The second path suggests that the politics of humans be melted into the politics of the relations of humans to the non-human: all are one. In a democratic constitutional state, no power relation can be absolute. A process of mediation and pondering of interests must be imposed by law to all power relationships. A law of intersubjective relations already exists. It is now time to conceive of a legal system that encompasses the power relations between objects and subjects. That system of law must primarily insert a moment of reflection, temporization and mediation in our behavior vis-à-vis the objects.

The third and last path inspires a fundamental reconsideration of environmental law. If the *Grands Partages* between nature and culture have been dismantled, and if all of our actions are seen to have an ecological dimension, the limited framework of the current environmental law may not be amenable to effectively and fundamentally rethink our legal relations with the non-human. It is equally doubtful whether the system of law can absorb truth-pluralism and the implications of the latter on our conception of the world, man and their (inter)relations, within the scope of an environmental law that lacks autonomy. These doubts suggest that the process of reflection had better address a more fundamental level of law, that could legally conceive the ecological dimension of *all* of our actions. This third and last path consequently insists on the necessity to pass the concepts that have been advanced as possible foundations for an environmental law through the sieve of truth-pluralism. The conjunction of a relational law and a scientific practice that is stamped by truth-pluralism, may be expected to generate fundamental shifts in the basic conception of environmental law, and, more generally, alter the legal relations between men and nature. A varied series of concepts must then be evaluated that appear to occupy an intermediate position between a purely utilitarian and anthropocentrical plea to leave man be the king of things, a plea that is copied from the Cartesian *maîtrise et possession* [15] on the one hand, and on the other hand, the theses of deep ecology, according to which nature, its animals, plants and other objects should accede to the status of legal subject (reducing man to just a another slavish particle of the 'given natural order') [43, 45]. A long term exercise must be carried out in the near future. In the scope of that exercise, the conclusions of truth-pluralism and the design of a relational legal practice will serve as a basis to evaluate such fundamental concepts as, amongst others, the 'common patrimony of humankind' [46], the 'responsibility towards future generations' [7, 29] and 'sustainable development' [47, 48], 'usufruct' conceived as an encompassing legal figure [51], a 'fundamental right to (the protection of) a healthy environment' [5, 6] and, finally, the idea of a 'natural contract' [52]. These evaluations and comparisons must give rise to still more elements that can help answer the question whether, all in all, 'environmental law' is the best instrument to realize the goals it pretends to realize.

Serge Gutwirth
Brussels Free University,
Center for the Interaction of Law and Technology,
Brussels, Belgium
e-mail: slgutwir@vub.ac.be

ACKNOWLEDGEMENTS

Thanks to An Vranckx for editing this article.

REFERENCES

[1]Often, as Serres shows, the history of the sciences is reduced to the analysis of the decisions of 'knowledge courts'. These courts decide whether the initial arguments ('les causes') can claim scientific recognition, truthfulness and can accede to status of fact ('les choses'): "Autrement dit, la connaissance scientifique résulte du passage qui fait de la cause une chose et de celle-ci celle-là, par où un fait devient un droit et inversément" [52a] and [53a]. From an etymological point of view the French words 'cause' and 'chose' have the same origin. In Latin the word *res* can be used in both senses, as is also the case for the Dutch word 'zaak'. Moreover, in Dutch 'geding' (controversy) and 'ding' (thing) exemplify a common etymological root.
[2] Indeed, as Latour writes, "les deux sens du mot représentation ne sont pas si éloignés que l'on croit dans les deux ordres du politique et du scientifique. Dans les deux cas, il y a des porte-parole qui diffèrent de leurs mandants. Dans

les deux cas il s'agit du remplacement du silence des mandants par la parole des mandatés. Dans les deux cas, les controverses sont réglées provisoirement par une suite d'épreuves où se met en scène la fidélité des mandatés à ce qu'auraient pu dire les mandants, la sûreté de leurs autorisations à parler, la légitimité des liens qu'ils maintiennent avec leur bases. Dans les deux cas, enfin, le résultat des épreuves se trouve lui-même stabilisé par un déplacement vers d'autres formes d'institution, d'appareillages ou de dispositifs. La querelle épistémologique entre réalisme et relativisme forme l'exact pendant de la querelle politicologique entre représentation directe et indirecte" [36a].

[3] Serious attention should consequently be paid to "le scandale des scientifiques lorsqu'on ramène leur prétentions à l'objectivité à un 'folklore particulier', susceptible du même type d'analyse que les folklores des autres pratiques humaines (...) la singularité des sciences n'a pas besoin d'être niée pour devenir discutable. Afin de faire des scientifiques des acteurs comme les autres dans la vie de la cité (...), il n'est pas nécessaire de décrire leur pratique comme semblable à toutes les autres"; [56b].

[4] "(...) il ne s'agit plus de vaincre le pouvoir de la fiction, *il s'agit toujours de mettre à l'épreuve*, de soumettre les raisons que nous inventons à un tiers susceptible de les mettre en risque"; [56c] and: "Seul le travail des scientifiques ensemble, producteur de controverses, seuls les travaux qui acceptent le défi d'avoir à *résister* aux autres, à *intéresser* les autres, à tenter de *modifier les degrés de liberté* qui sont ouverts aux autres, peuvent êtres caractérisés comme scientifiques" [55a]. See also Michel Callon: "Le secret de la science tient tout entier dans ces deux impératifs entre lesquels le chercheur ne peut pas choisir: éprouver la résistance d'une proposition, mesurer l'intéressement qu'elle suscite. Cette double exigence définit l'historicité propre de la connaissance scientifique et son horizon indépassable" [9a].

[5] **Boyle, A. & Anderson, M.** (Ed.), *Human rights approaches to environmental protection*, Oxford, Clarendon, 1996, 313 p.

[6] **Brandl, E. & Bungert, H.**, "Constitutional entrenchment of environmental protection: a comparative analysis of experiences abroad", *Harvard environmental law review*, **Vol. 16**, nr. 1, 1992, pp. 1-99.

[7] **Brown Weiss, E.**, *In Fairness to Future Generations: International Law, Common Patrimony, and Intergenerational Equity*, Tokio/New York, The United Nations University/Transnational Publishers, Inc., 1989, 385 p.

[8] **Callon, M. & Rip, A.**, "Humains et non-humains: morale d'une coexistence" in: **Theys, J. & Kalaora, B.** (Dir.), *La terre outragée. Les experts sont formels!*, Paris, Editions Autrement, Série Mutations/Sciences en société **nr. 1**, 1992, pp. 140-156.

[9] **Callon, M.** (Dir.), *La science et ses réseaux. Genèse et circulation des faits scientifiques*, Paris, La Découverte, 1989, 214 p.

[9a] *ibid.*, pp. 32-33.

[10] **Collart Dutilleul, F. & Romi, R.**, "Propriété privée et protection de l'environnement", *A.J.D.A.*, 20 septembre 1994, pp. 571-582.

[11] **Delmas-Marty, M.**, *Pour un droit commun*, Le Seuil/La librairie du XXe siècle, Paris, 1994, 306 p.

[12] **Dosse, Fr.**, *L'empire du sens. L'humanisation des sciences humaines*, Paris, La découverte, 1995, 432 p.

[12a] *ibid.*, p. 129.

[13] **Edelman, B. & Hermitte, M.-A.** (Ed.), *L'homme, la nature et le droit*, Paris, Bourgois, 1988, 392 p.

[14] **Ewald, F.**, *L'Etat providence*, Paris, Grasset, 1986, 608 p.

[15] **Ferry, L.**, *Le nouvel ordre écologique. L'arbre, l'animal et l'homme*, Paris, Grasset, 275 p.

[16] **Foqué, R. & 't Hart, A.C.**, *Instrumentaliteit en rechtsbescherming. Grondslagen van een strafrechtelijke waardendiscussie*, Arnhem/Antwerpen, Gouda Quint/Kluwer, 1990, 501 p.

[17] **Foqué, R.**, *De ruimte van het recht*, Arnhem, Gouda Quint, 1992, 44 p.

[18] **Foqué, R.**, "Le droit et sa fin politique", *Tijdschrift voor de studie van de verlichting en het vrije denken*, 1988/2-4, pp. 145-159.

[19] **Gutwirth, S.**, *Waarheidsaanspraken in recht en wetenschap. Een onderzoek naar de verhouding tussen recht en wetenschap met bijzondere illustraties uit het informaticarecht*, VUBPress, Brussels, 1993, 846 p.

[19a] *ibid.*, p. 20.

[20] **Gutwirth, S.**, "Autour du contrat naturel" in: **Gérard, Ph., Ost, F. & van de Kerchove, M.** (Eds.), *Images et usages de la nature en droit*, Bruxelles, Publications des Facultés universitaires Saint-Louis Bruxelles, 1993, pp. 75-131.

[21] **Gutwirth, S.**, "Mens, natuur en recht. Grondslagentheoretische bedenkingen bij de verhoudingen tussen rechtssubjecten en rechtsobjecten" in: **Gutwirth, S.** (ed.), *Milieu rechtgezet? Een bezinning over de grondslagen en de toepassing van het milieurecht*, *Tegenspraak Cahier nr. 15*, Gent, Mys & Breesch, 1994, pp. 1-31.

[22] **Gutwirth, S.**, "Het harlekijnspak en het vale ezelsvel. Individuele vrijheden in de greep van het recht op zelfbeschikking van volken, het nationalisme en andere aanspraken op een collectieve identiteit" in: **Sybesma, N. & Van Bellingen, J.** (Eds.), *Naar een nieuwe interpretatie van het recht op zelfbeschikking?*, Brussel, VUBPress, 1995, pp. 53-88.

[23] **Gutwirth, S.**, "Sciences et droit de l'environnement" in: **Ost, F. & Gutwirth, S.** (Eds), *Quel avenir pour le droit de l'environnement?*,

Brussel, Publications de Facultés universitaires Saint-Louis/VUBpress, 1996, pp. 21-42.

[24] **Gutwirth, S. & Naim-Gesbert, E.**, "Science et droit de l'environnement: réflexions pour le cadre conceptuel du pluralisme de vérités", *Revue interdisciplinaire d'études juridiques*, 1995, **34**, pp. 33-98.

[25] **'t Hart, A.C.**, *Recht als schild van Perseus. Voordrachten over strafrechtstheorie*, Leerstoel Theodore Verhaegen 1989-1990 - Vrije Universiteit Brussel, Arnhem/Antwerpen, Gouda Quint/Kluwer, 1991, 191 p.

[26] **'t Hart, A.C.**, *Openbaar ministerie en rechtshandhaving*, Arnhem, Gouda Quint, 1994, 402 p.

[27] **'t Hart, A.C.**, *Mensenwerk? Over rechtsbegrip en mensbeeld in het strafrecht van de democratische rechtsstaat*, Mededelingen van de Koninklijke Nederlandse Akademie van Wetenschappen, Afdeling Letterkunde, Noord-Hollandsche, Amsterdam, 1995, Nieuwe reeks, **Deel 58**, no. 4, 40 p.

[28] **'t Hart, A.C.**, *De meerwaarde van het strafrecht*, Sdu, Den Haag, 1997, 318 p.

[29] **Jonas, H.**, *Le principe responsabilité*, Paris, Cerf, 1991, 336 p.

[30] **Latour, B.**, *Les microbes. Guerre et paix suivi de irréductions*, Paris, Métaillé, 1984, 282 p.

[31] **Latour, B.**, *Wetenschap in actie*, Amsterdam, Bert Bakker, 1988, 348 p. (Original title: Science in action)

[32] **Latour, B.**, "Pasteur et Pouchet: hétérogenèse de l'histoire des sciences"in: **Serres, M.** (Ed.), *Eléments d'histoire des sciences*, Paris, Bordas, 1989, pp. 423-445.

[33] **Latour, B.**, *Nous n'avons jamais été modernes. Essai d'anthropologie symétrique*, Paris, La Découverte, 1991, 211 p.

[34] **Latour, B.**, *Aramis ou l'amour des techniques*, Paris, La Découverte, 1992, 241 p.

[35] **Latour, B.**, "Arrachement ou attachement à la nature", *Ecologie politique*, **nr. 5**, Hiver 1993, pp. 15-26.

[36] **Latour, B.**, "Esquisse d'un Parlement des choses", *Ecologie Politique*, numéro 10, été 1994, pp. 97-115.

[36a] *ibid.*, p. 107.

[37] **Latour, B.**, "Moderniser ou écologiser? A la recherche de la "septième" cité", *Ecologie Poltique*, **no. 13**, Printemps 1995, pp. 5-27.

[38] **Latour, B., Schwartz, C. & Charvolin, F.**, "Crise des environnements: défi aux sciences humaines", *Futur Antérieur*, **6**, Eté 1991, pp. 28-56.

[39] **Latour, B. & Woolgar, S.**, *La vie de laboratoire. La production de faits scientifiques*, Paris, La Découverte, 1988, 300 p.

[40] **Lascoumes, P.**, *L'éco-pouvoir*, Paris, La Découverte, 1994, 318 p.
[41] **Morin, E.**, "Pour une pensée écologisée" in: **Theys, J. & Kalaora, B.** (Dir.), *La terre outragée. Les experts sont formels!*, Paris, Editions Autrement, Série Mutations/Sciences en société **no. 1**, 1992, pp. 66-77.
[42] **Morin, E. & Kern, A.B.**, *Terre-Patrie*, Paris, Seuil, 1993, 217 p.
[43] **Naess, A.**, "The shallow and the deep, long-range ecology movement. A summary", *Inquiry*, **vol. 16**, 1973, pp. 95-100.
[44] **Naim-Gesbert, E.**, "Expertise scientifique et droit de l'environnement" in: **Ost, F. & Gutwirth, S.** (Eds), *Quel avenir pour le droit de l'environnement?*, Brussel, Publications de Facultés universitaires Saint-Louis/VUBpress, 1996, pp. 43-88.
[45] **Nash, R.**, *The rights of nature. A history of environmental ethics*, Madison, Univ. of Wisconsin Press, 1989, 291 p.
[46] **Ost, F.**, *La nature hors la loi. L'écologie à l'épreuve du droit*, Paris, La Découverte, 1995, 346 p.
[47] **Pallemaerts, M.**, "La conférence de Rio: grandeur ou décadence du droit international de l'environnement", *Revue Belge de Droit International*, 1995/1, pp. 175-223.
[48] **Pallemaerts, M.**, "International environmental law from Stockholm to Rio: Back to the future?", *Review of International and European Community Environmental Law*, 1992, **Vol 1**, nr. 3, pp. 254-266.
[49] **Prigogine, I. & Stengers, I.**, *La nouvelle alliance. Métamorphose de la science.*, Paris, Gallimard/Essais **26**, 1979, 439 p.
[50] **Poincaré, H.**, "Le libre examen en matière scientifique", *L'université de Bruxelles. LXXVe anniversaire de la fondation. Relation des fêtes. Novembre 1909*, Bruxelles, Weissenbruch, 1910, pp. 97-106.
[51a] *ibid.*, p. 98.
[52] **Rémond-Gouilloud, M.**, *Du droit de détruire. Essai sur le droit de l'environnement*, Paris, PUF, 1989, 304 p.
[52] **Serres, M.**, *Le contrat naturel*, Paris, François Bourin, 1990, 191 p.
[52a] *ibid*, p. 43.
[52b] *ibid*, pp. 26-45.
[53] **Serres, M.**, *Statues*, Paris, Bourin, 1987, 346 p.
[53a] *ibid*, pp. 110-111.
[54] **Serres, M.**, *Hermès III. La traduction*, Paris, Minuit, 1974, 269 p.
[55] **Stengers, I.**, "Les "nouvelles sciences" modèle ou défi", *Review of the Fernand Braudel Centre*, winter 1992 (theme: The "new science" and the historical social sciences), pp. 91-112.
[55a] *ibid.*, p. 108.
[56] **Stengers, I.**, *L'invention des sciences modernes*, Paris, La découverte, 1993, 210 p.
[56a] *ibid.*, pp. 76-80.

[56b] *ibid.*, p. 70.
[56c] *ibid.*, p. 151.
[57] **Touraine, A.**, *Qu'est-ce la démocratie?*, Paris, Fayard, 1994, 297 p.

WILLY WEYNS

STRENGTH AND WEAKNESS (*SWOT*) ANALYSIS OF HUMANKIND TO OVERCOME THE GLOBAL ECOLOGICAL CHALLENGES

INTRODUCTION

At the treshold of the 21st century, we live in a world that is changing at ever higher pace. In the same time this changing world is under pressure of unprecedented ecological and cultural (in the broad, anthropological sense) challenges. According to many eco-philosophers and cultural philosophers, this state of affairs might be due to a great extent to mechanisms that have proven to be so successful at micro-level, but that seem to be ineffective or counter-productive at the global scale. I am pointing, e.g., at the competitive market mechanisms, that are forcing private corporations to increase very effectively and continuously their material and energy throughput.

SWOT-Analysis

In the state of urgency that has been the result of all this, I refuse to speak to you in a virtual reality. In spite of the abovementioned presumptions, I would like to invite you to a Magritte styled approach, by briefly investigating what a successful management technique i.c., the *SWOT*-analysis, could yield at the global level. This is a rather surrealistic approach, because a *SWOT*-analysis has precisely developed in the market mechanism that we suspected to be at the origin of most global problems. This management technique is composed of a listing and analysis of strengths, weaknesses, opportunities and threats to the company's development. My endeavour at the global level is composed of the same four parts, and in each part I will limit myself to four crucial points, which yields the following overall picture:

STRENGTHS

* Human consciousness, of which the originally western **scientific enterprise** and more specifically the monitoring of environmental criticality is of particular relevance here: It is thanks to this scientific approach

© 1999 *Kluwer Academic Publishers. Printed in The Netherlands.*

TABLE

Realm	*Strength*	*Weakness*	*Opportunity*	*Threat*
Scientific	Environmental monitoring	Fragmentation	Citizens Tribunal	Scientism
Technological	Enhanced impact	Material throughput too high	Solar future	Technocracy
World view & attitudes	Cultural evolution	Feeble willingness-to-pay, abstractness	Paradigm change, well-being	Consumerism, mimetism
Political	Pluralism	Short-term planning	Civil society, World citizenship	Time lags, N/S gap

that almost invisible and upcoming threats are discovered and assessed and that a large consensus is reached about possible scenarios for action. We think of radio-active materials, the so-called hole in the stratospheric ozone layer, the anthropogenic greenhouse warming and so on. We point at emerging action plans such as developed by the International Panel for Climatic Change (IPCC) and Agenda 21 (UNCED).

∗ **Technological means**, enhancing the real and potential impact of humankind on nature, mostly in the richer industrial countries. This tremendous power of humans, with both its intended, constructive (e.g., material culture, infrastructures, huge dams, ...) or destructive (e.g., nuclear, biological and chemical mass destruction weapon systems) impacts and unintended side effects (routine pollutions, accidents type Exxon Valdez, Bhopal and Chernobyl), is listed here as a strength, because the magnitude of the present 'technometabolism' [6] (7,3 tons of gravel and stones and 20 tons of CO_2 per person yearly) seems to be matched with the planetary scale of the challenges we face today.

∗ The human aptitude of a **cultural evolution**, being much quicker than the bio-evolution, with a substantially enhanced global adaptability of humankind to changing environments as a result [5]. This higher gear is of paramount importance, because time seems to be the crux of the matter, as will be notified further.

∗ **Cultural diversity** and existence of non-violent counter-currents and dissident opinions within numerous cultural settings. This tolerance that finds its latest expression in 'constructive postmodernism', favours the occurrence of cultural paradigm shifts through opening the powerful communication canals to the expression of novel ideas and metaphors. In this way, just as biodiversity, cultural diversity is credited a high survival value for humankind [1] [4] [23].

This forcibly limited list feeds some speculations of the salvatory potentials of human **spare brain capacity** to develop in the future, but we have to counter this optimistic view with a consideration of four weaknesses now:

WEAKNESSES

> *'Of all man's miseries, the bitterest is, to know so much, and to be impotent to act'*
>
> (Herodotus, 5th Century B.C.)

* The transition from the awareness (i.e., of a global environmental crisis) to the changing of behaviour, of lifestyle. In other words the **limited willingness to pay**. At the Earth Summit (UNCED) at Rio de Janeiro in 1992, those who would have to pay the most (The USA and the OPEC countries) were outspoken opponents against a strong and binding Climate Convention. The rich are not willing to decrease their excessive material welfare, whilst the poor overwhelmingly don't refrain from copying this same wasteful lifestyle. In the first case theories of compensatory consumerism in the postmodern condition [7], and of mimetic desire in the second [12] are attempts to explain these powerful phenomena. But can they massively be curbed in time?

* **Fragmentation** in Science and extreme dualisms on the personal level (the consumer in each of us opposed to the citizen) and in society as a whole. M. Sagoff [22] and M. Jacobs [16] i.a. have explored the reduction of the political sphere to a mere expression of consumer preferences in the ever expanding market. The dominant myth of an economy that is apart from or above the environment, leading to its logical extreme in the dream of life in a virtual reality. Instantaneous transfer of huge speculative funds in cyberspace: a casino economy decoupled from the real economy on which production and jobs depend. This brings us to the following point:

* **Complexity**, abstractness, unpredictability of the evolution of global systems (and ecological problems) in combination with the bias of human mind to simplify and narrow the problem to single-issue themes, as is particularly visible in the case of political activism [26]. Many authors have largely developed the point that the specific character of environmental challenges (complexity, uncertainty, high stakes) demand a new type of rational approach including a new articulation between scientists and informed citizens [11] [27].

* The **lack of adequate feedback** mechanisms, as denounced in various reports to the Club of Rome [17]. Reactions in the political and economic sphere, such as the internalization of environmental costs are bound to be always too little and too late.

The common denominator of those weaknesses could be the dominant value-system ('**pathologies of affluence**') felt around the world, being totally unadapted to the prevailing global challenges in the real world. To this objectively existing situation in our world we turn now:

OPPORTUNITIES

* The global crisis, due both to environmental and societal (or broadly cultural) challenges could give rise to a **unifying agenda** (type Agenda 21 of the United Nations Conference on Environment and Development—UNCED—at Rio de Janeiro, 1992 [28]). In fact, in this global age, there is no solution to the global challenges without a massive world-wide cooperation, and such type of adhesion has no chances to be obtained without taking into account fundamental socio-political issues such as the ever widening development gap, international trade, debt burdens and so on [13].

* The need for unity could foster inter-cultural negotiations among emerging cultural-politico-geographic aggregates (the West, the Islamitic world, China, ...[9] [20]). Due to the extreme complexity and intertwining of the environmental socio-economic and cultural problems at stake, those negotiations will be much more difficult to bring to a good end than the historical disarmament talks. Notwithstanding they could possibly foster an **overlapping consensus** about a limited but fundamental global strategy. I have introduced this line of thought already in the editorial of this green book [19] [21].

* **New social** (the global civil society, nonviolent action models, a new articulation between hard scientific expertise and institutions expressing

societal values [31]) **and technological means** (cybernetics), might be the core of an adequate feedback mechanism in construction. Citizen's Tribunals, placing scientific experts inside a contradictory type of public Court procedure [24], massive nonviolent actions, citizens on the other side of the globe reacting in real time on political events, are but a few examples taken at random of what could become a new, postmodern citizenship [2] [10] [14] [15] [18] [29] [30].

∗ The global energy crisis might favour a **solar future**. The end of the geologically accumulated fossil fuels (except for coal) coming in view might facilitate the economic and institutional transition towards already existing solar technologies, directly capturing the solar radiations reaching the planet. Such methods avoid the full cycle of petrification and its consequent raise of global entropy.

Those opportunities are foundational of a hope that the global crisis at hand be an instigator of a constructive postmodern readjustment. But even in such an optimistic perspective, is there a chance that this reaction will come in time?

THREATS

∗ Important **time lags** (of decennia) exist in global systems (e.g., the greenhouse warming) and are in contrast with societal systems (democracies in the first place) focusing on the short-term;

∗ Severe **ecosystem breakdown** cannot be excluded: loss of primary productivity of continental shelves (humankind now already is appropriating a substantial 40% of continental bioproductivity), irreversible loss of genetic material (tropical rainforest destruction [32], unsustainable energy consumption, technometabolism in excess of carrying capacity are alarming signals indeed...[3] [6] [25])

∗ The increasing social, political and economic **gap between the rich** (dominantly, but not exclusively in the northern hemisphere) **and the poor:** wages relate grossly 23 to 1 [33]—If all inhabitants of the world started consuming as much energy or driving their own private car, as most of us do today, this would be absolutely catastrophic from an ecological point of view! But every citizen of the world being entitled the same rights, why should they not?

∗ Technological way of being and thinking (**'technical fix'- mentality**) absolutely dominating, instead of technology being steered by societal and cultural goals.

Globally speaking, there are reasons enough to fear the ecosystem earth self-readjusting at a level offering sensibly **lower development opportunities** both to humans and to other (higher) organisms.

COMING TO A CONCLUSION ...

The transformation of the earth's surface always has been and will be an inescapable concomitant of human societies. The question here raised is whether the pattern of interplay could become such that a massive ecosystem breakdown and readjustment at a substantial lower level of complexity and higher entropy could be avoided. The central question might be whether human societies are capable of becoming sustainable, of adapting to limits. The precautionary principle [8] dictates, and there are so many indications by now anyway, that biological systems can only be hoped to continue to function in the interests of humanity itself, only by limiting the global ecological load of the human species on earth.

This inevitably brings in the determinism↔free will question: can ideas help to change the hard realities of economics and politics? If we didn't believe in at least a dialectical relationship between both, we would not stand here, exposing insights, defending ideas and proposing an action-oriented research program. The diagnosis pointing among others at an outdated value-system, how can values and norms be informed by factual scientific knowledge, be publicly discussed and made revisable?

My last word will be about the need for powerful and adequate new metaphors. We have to fight the metaphor of unlimited material growth, but we cannot tell our children not to grow. If we want humankind to re-find its roots in nature, not outside, we are in need of the right metaphor. *Einstein meets Magritte* is a strong metaphor for this conference. Hence we have to be creative as many Einsteins and Magrittes having met, from now on!

Willy Weyns
VUB—HUCO—CIEPAC
Wildestraat 14, B-9961 Assenede, Belgium
weyns.roscam@glo.be

REFERENCES

[1] **Apostel, L.**, *Wereldbeelden, ontologie en ethiek*, Humanistisch Verbond, Antwerp, 1994.

[2] **Axelrod, R.**, *De evolutie van de samenwerking*. Contact, Amsterdam, 1990.

[3] **Arrow,K et al.**, "Economic growth, carrying capacity and the environment", *Science*, **268**, 1995, pp. 520-21.

[4] **Batens, D.**, "Rationality and ethical rationality", *Philosophica*, **22**, 1978, pp. 23-45.

[5] **Boyd, R. & Richerson, P.**, *Culture & the evolutionary process*. U.C.P., Chicago, 1985.

[6] **Boyden, S.**, "Human ecology & societal change", *Journ. Hum. Ecol. I*, **1**, 1990, pp. 7-19.

[7] **Brennan, Th.**, "The Age of Paranoia". In: **Kunneman, H. & de Vries, H.** (eds), *Enlightenments*, Kok, Kampen, 1993.

[8] **Dovers, S. & Handmer, J.**, "Ignorance, the precautionary principle and sustainability", *Ambio XXIV*, **2**, 1995, pp. 92ff.

[9] **Fischer, J. & Ury**, *Getting Yes*, Penguin, London, 1981.

[10] **Fox, W.**, *Toward a transpersonal ecology*, Shambhala, Boston, 1990.

[11] **Funtowics, S. & Rawetz, J.**, "Science for the Post-normal Age". In: **Westra, L. & Lemons, J.** (eds), *Perspectives on Ecological Integrity*, Kluwer Ac. Pub., Dordrecht, 1995.

[12] **Girard, R.**, *Des choses cachées depuis la fondation du monde*, Grasset, Paris, 1978.

[13] **Goodland, R.**, "The urgent need for ecosystem integrity and ethics to support environmental sustainability". In: **Westra, L. & Lemons, J.** (eds), *op.cit.*, 1995.

[14] **Guattari, F.**, *Les trois écologies*, Galilée, Paris, 1989.

[15] **Jacobs, M.**, "Sustainable development and deep ecology: an analysis of competing traditions", *Env. Manag. IIXX*, **4**, 1994, pp. 477-88.

[16] **Jacobs, M.**, "Environmental valuation, deliberative democracy and public decision making institutions", In: **Foster, J.** (ed), *Valuing nature*, Routledge, London, 1996.

[17] **Meadows, D. et al.**, *Beyond the limits*, Earthscan, London, 1991.

[18] **Meyer-Abich, K.**, *Revolution for nature*, White Horse Press, Ca, 1993.

[19] **Naess, A.**, *Ecology, community and lifestyle*, C.U.P., Cambridge, 1989.

[20] **Pinxten, R.**, *Culturen sterven langzaam*, Hadewijch, Antwerpen, 1994.

[21] **Rawls, J.**, *Political Liberalism*, New York, 1993.

[22] **Sagoff, M.**, *The Economy of the Earth*, C.U.P., Cambridge, 1989.
[23] **Sattler, R.**, *Biophilosophy*, Springer, Berlin, 1986.
[24] **Schrader-Frechette, K.**, *Science, policy, ethics and economic methodology*, Reidel, Dordrecht, 1985.
[25] **Simmons, I.**, *Changing the face of the earth*, Basil Blackwell, Oxford, 1989.
[26] **Straus, D.**, "Thinking the unthinkable to avoid the unthinkable crisis", *Hum. Ecol. Rev. I* , **1**, 1994.
[27] **Tuler, S. & Webler, T.**, "Process evaluation for Discursive Decision Making in Environmental and Risk Policy", *Human Ecology Rev.*, **2**, 1995, pp. 62-71.
[28] **UNCED**, *Agenda 21.* UN: New-York, 1992.
[29] **von Weizsäcker, E.**, *Earth politics*, Zed, London, 1994.
[30] **Weyns, W.**, *Het gras onder onze voeten.* VUBPress, Brussels, 1995.
[31] **Weyns, W.**, "Scientific objectivity and worldviews in impact studies", *Int. Sympos.E.I.A. Water Management, Bruges May 15-17, 1995 (Book of abstracts).* GTE/VMM/AMINAL, Brussels, 1995.
[32] **Wilson, E.**, *The Diversity of Life*, Harvard University Press, 1992.
[33] **WORLDBANK**, *Development Report 1995*, 1996.

JAN FENNEMA

"AND YOU, MY MOUNTAIN, WILL YOU NEVER WALK TOWARDS ME?"

INTRODUCTION

The title of this text, the subject matter of a lecture, is a phrase, a motto, taken from a book by the Norwegian philosopher Arne Naess. It is a poetic title, and an enigmatic one as well. In a special sense, it can be considered the introductory phrase of Naess' monograph *Ecology, Community and Lifestyle* [1], a book that is very much worth reading. The title may stimulate the reader to ask questions, for instance, about the meaning of this introductory phrase, about what it intends to convey. Besides, as the text leads us to contemplate the relation of ecology to theology, and vice versa, without doubt other questions will follow. I will try to answer such questions, but the reader is warned: the answers will doubtless give rise to further questions.

ECOPHILOSOPHY AND ECOSOPHY

This century, and not least this decade, is full of science with ever increasing sophistication and universal impact. And it is equally full of a pervading sophisticated meta-science [2], by which I mean, for instance, theories of cosmic space and time and also theories concerning the evolution of species, the development of the human being in particular; cosmology, 'Big Bang', and molecular biology should be mentioned in this respect. Reflecting on these aspects of present day global culture, which are well-known, awareness of the presuppositions of our ways of thinking is stimulated. The development of scientific culture as sketched reveals a maxim: *scientia vincere tenebras*, that is, 'to conquer by means of science what remains obscure'; recently I learned that the Brussels Vrije Universiteit—Free University—carries these words on its coat of arms.

'Scientia vincere tenebras' is a typical Enlightenment phrase, which reminds us of the aspiration of that epoch of our culture to bring light into the darkness, more specifically, to shed light on things that were obscure—or that were thought to remain obscure—because of the dogmatic views of regnant authorities, church authorities mostly at that time. What does the word 'science' mean in this context? Naturally it

© 1999 *Kluwer Academic Publishers. Printed in The Netherlands.*

means careful experimentation, a rational analysis and consequent synthesis, and, of course, a belief in the reliability of this procedure when seeking after truth. And what is understood by 'truth' in this connection? The term 'truth' implies, among other things, a coherent knowledge of the world we are imbedded in, that is insight into the why and how of things, including ourselves; it leads to applying what we have thus learnt, possibly in order to make a better world. Arne Naess in his monograph reflects on the world as it is today: maybe, he writes, we intend to make a better world, but in fact we are facing catastrophe. For, quoting his words, 'an exponentially increasing and partially or totally irreversible environmental deterioration or devastation [is] perpetuated through firmly established ways of production and consumption and a lack of adequate policies regarding human population increase'. Naess and his collaborators are convinced that quick-fix solutions, although doubtless needed also, are not sufficient if we want to solve our global problems. For, at bottom, there are false conceptions of reality at work that determine the direction of our civilization, and those conceptions must be criticized in order to avoid final destruction. Naess subsequently explains that 'new types of societies and communities' are needed in which 'being together with other living beings is more important than exploiting or killing them'. As is well-known, a reflection of this kind on the condition of humankind is generally called 'ecophilosophy'.

But to Naess and his collaborators ecophilosophy is not just a system, or a paradigm, that is applied, not just their personal paradigm. For their way of thinking presupposes *wisdom*! Systems and paradigms are intellectual constructs; they are instruments of heuristic value, conceptual frameworks that function when a complete theory is not yet available. More specifically, a system may (here) be considered a method of organization and a paradigm an example formally serving as a pattern. In contradistinction, wisdom is first and foremost existential. Following the definitions given in dictionaries [3], '[Wisdom] cannot be learned or understood as a theory' as it is 'not what someone possesses but what someone is'. Naess and his collaborators cruelly experienced the existential core of their ecophilosophy when one of their friends died under an avalanche that suddenly rushed down a mountain. A tragedy that was articulated, cried out by them in the words: 'And you, my mountain, will you never walk towards me?'[4]. These words, generated by immediate experience, constitute, so to speak, the opening phrase of the book which was written at that time. This outcry of extreme despair confronts us with the root of their predisposition, which is conviviality with Nature, a togetherness with what is ours—because we are Nature also ourselves—and a togetherness with what is different, distinct from what we are. Clearly, a togetherness with what was so strikingly different from the group of human beings they were themselves: face-to-face with a mountain that impressed them; that is, face-to-face with the numen, the numinous of a mountain that killed one of the group members. In other words, their ecophilosophy fed upon a wisdom that implied being *one* with what at the same time was not (the) self but (the) *other* intrinsically. Naess calls this wisdom 'ecosophy'. And in his monograph he is quite explicit with regard to the notion '(religious) humility in the face of Nature'. Formally this wisdom, ecosophy, implies a search for a balance between what is understood as (the) self and what is considered as (the) other. Consequently in Naess' view, any ecosophy is always a particular ecosophy rooted in the experiences of a particular person, who has searched for a personal balance. Hence, what matters is not the singular, 'ecosophy', but the plural, that is 'ecosophies'.

Is ecosophy something characteristic of our time with its threat of ecological disaster? Or is ecosophy a wisdom of old, a wisdom of which we are reminded by impending global catastrophe, but to which we might return if we are sufficiently prudent? It is worth-while attempting a sketchy answer to this question. The need for ecosophy—here and hereafter shorthand for the plural 'ecosophies'—does not stand isolated. Indeed, to mention just a few examples, the present need for ecosophy, and the continuing debate concerning the relation between science and religion since the development of European scientific culture, and the age-old dispute between philosophy and theology have a fundamental value in common. In my opinion they all point to the same basic issue: the status of humankind as a being in quest of wisdom to find a proper balance between 'self' and 'other' [5]. Thus, the fundamental problems are the status of humankind and the consequent dialectics between a human being that is considered self-sufficient and a human being to whom existence is and remains a gift, a grace received from what is not-self but other, from beyond the horizon of private space-time [6]. In (contentiously) traditional terms, it is the dialectical relationship of the *human* and the *divine* in the human being as prefigured already in the debates of the early centuries of Christianity, in particular with reference to the so-called Chalcedonian controversies. What follows is an attempt to clarify this ultimately mysterious statement.

THE HUMAN INDIVIDUAL: AN ECCENTRIC BEING

At the present time, in the opinion of many scholars, theology belongs to the realm of the private, to coteries maybe, and it has a bearing supposedly only on their private perspectives and convictions. But in the European context, theology, philosophy and the sciences cannot be rigorously separated, for all have played an important role in the development of European culture. Since the beginning of the common era, philosophy, science and theology—the last term in fact referring to Jewish and Christian thinking predominantly—have existed side by side, at times in harmony, trying to develop some comprehensive view, or at variance with each other and intensively interacting in a continued power struggle. Thus, the emergence of any monomania has never remained uncontested, save for short durations and in restricted domains. For, no doubt, a system of thought that closes in upon itself and becomes rigid is, and always will be, challenged in due course in order to renew its content or even to be abandoned altogether. Consequently, philosophical, scientific and theological thinking have continued to develop. In fact, far from serving exclusively the restricted interests of limited groups, philosophy, science and theology possess worldwide dimensions. They *all* have a bearing on our understanding of the universe; they all direct our perspectives and help to determine our attitudes. Indeed, European civilization has been distinctively moulded thereby. And ecosophy may represent the summit of its achievement, decisively settling its scope in our hectic and confusing present.

In my opinion the wisdom of 20th century ecosophy finds support in the debates of early Christianity, though on matters that appear to serve entirely different interests. A few remarks may suffice to indicate what I have in mind. The concern of the debates preceding, during and following the Council of Chalcedon, a town in Asia Minor, in the year 451, was the understanding of what were called the human and divine natures in Jesus Christ, who was believed to be one person. These Christological debates crucially anticipated present day anthropology [7] in so far as the latter attempts to reflect on the human individual as a being which is eccentric, that is, as a structure which is, so to speak, *energetically open.* Doubtless this statement, enigmatic prima facie, is in need of some elucidation with regard to the content of the Christological debates and also with regard to anthropology. We begin with the latter.

To avoid misunderstanding it is straight off emphasized that the connotation of the adverb 'energetically' is not, at any rate not in the first place, so-called physical reality, the world of physics. Evidently, the human body as an open system draws upon energetic exchanges with the environment, and in this sense the human system depends on the global system; it is energetically open in an obvious way. But this meaning of 'energetically open' is not intended. If the human individual is characterized as an eccentric being and an energetically open structure, the connotations are in fact primarily meant to be existential. In order to live and to be able to continue living, a human being is in need of more than what is required by the metabolism of the body. For the *expectation* to be able to live and survive is crucial. And the basis of such expectation is *hope*. In my opinion the metabolism of the body and the resulting physical and psychical power/empowerment of a particular human being are not sufficient unless hope is already there [8]! In this way (an essential element of) hope is from outside one's private space-time, from beyond its horizon. And in this connection hope must be said *not* to be confined to the world of one's private space-time. Thus hope (essentially) belongs to another world. To some extent hope may be considered a fundamental attribute of the will to live, but more basically, it precedes the will; for it generates it. Clearly, whatever arguments can be adduced to justify the hope that is cherished and the expectation that results, they will never be conclusive. Thus the act of hoping is basic and cannot be ignored—though this is often forgotten in conditions of affluence. In principle, hoping is 'against one's better knowledge'—as the saying goes, which naturally reflects past experience—because in the act of hoping a boundary is transcended and proleptic knowledge created, that is knowledge concerning a realm that is out of reach, 'utopic'—which literally means that it has no place, cannot be localized indeed. Thus the existential core of the individual transcends a boundary, a horizon in space-time, which is tantamount to characterizing the human individual as eccentric. The statement that the human being is energetically open specifies more explicitly that the essential eccentricity of the individual being is a possible source of empowerment. Besides, as the locus of hoping reaches beyond the horizon of one's private space-time, it must be concluded that being is intrinsically a gift, a grace that is received, despite all efforts to master it.

In other words, adopting the terminology of present day anthropology, if the human individual is characterized as an eccentric being, it is held here that the individual is a truly human being only if she or he is open to receive her or his (future) being from a realm that is out of reach, beyond her or his personality; in this context the term 'personality' is to be understood in the naïve, unreflected sense of the total of characteristics which are thought to make up the human individual. Consequently, humankind's pretended selfsufficiency, briefly autonomy, for instance in 'making' the future, can be maintained only at a superficial level. This implies that, on closer inspection, humankind must definitely be regarded as heteronomous, which literally means that 'laws' are 'obeyed' that are *external* to the human individual; clearly, the thrust of the technical terms 'laws' and 'obeyed' is metaphorical. This statement on the autonomy and heteronomy of humankind suggests consequences that will appear in human behaviour. For it embraces humankind's status as an enduring (im)balance beween autonomy in appearance and heteronomy in truth.

The elucidation in the light of the Christological debates focuses on the extent these debates in the fifth century A.D. prefigure the discussion on the eccentric positionality [9] of the human being in our century. As may be known, the Christological debates considered the personhood of Jesus Christ, who was equally called 'Son of Man' and 'Son of God'. Could this belief be explained in terms that referred to distinctive qualities of his personhood? The conclusion of the debates was that two 'natures' were to be distinguished in the personhood of Jesus Christ, the 'human' and the 'divine' that—in a cryptic wording that has become famous—'did not mix or change and were not separable or divided'. For us who experience Western technological society and its values it is difficult to assess the true meaning of this formula. For all that, in my opinion, an analogy transpires and a tentative conclusion can be put forward. Assume that the 'human' in the debates refers to what in everyday speech is considered the naïvely human which as we believe is principally to master; assume moreover that the 'divine' in the debates refers to what transcends our mastery and whatever other human ability, being principally out of reach, as was/is the naïvely divine in everyday understanding. Then the outcome of the debates clearly suggests that a human individual is a being characterized by an eccentric positionality and a consequent structure that is intrinsically open [10].

Must this result be considered contentious, improper perhaps, and not at all trustworthy because it was obtained by a tour-de-force which implied distortion? Obviously, this is a rhetorical question which finds a clear answer by suggestion from its context. Hence, the outcome is that a heuristic device has been obtained, a device that enables one to relate that which is culturally distant in time but which may be less far apart on choosing the proper perspective. Its merit thus being the intimation of a possible cultural continuity, the time may now have come to proceed one step further.

ECOLOGY AND THEOLOGY

The result obtained in the preceding paragraphs clearly presupposes that the divine, as referred to in the Chalcedonian disputations, and the numinous mentioned in connection with the outcry, 'And you, my mountain, will you never walk towards me?', can be related. Evidently, this in turn suggests a speculation on the possible relation of theology to ecosophy or ecophilosophy, in short, to ecology, and vice versa.

In this respect, some will hold, in naïve terminology, that 'the world and God are one', a provocative tenet which leads to pantheism, and which may be supported by an appropriate natural theology, that is a theology constructed on the basis of a study of nature. Others reject the identity intimated in pantheistic doctrine, and confess, more subtly, a pan*en*theistic creed—briefly: 'the world is in God'—and develop a theology of nature, that is a reflection on nature in the light of an accepted theology.

The interpretation of the terms used in the two preceding sentences is of paramount importance. In my opinion, it is crucial for a human being whether one's existence is understood as something that is self-sufficient or accepted as a gift. In a similar vein the world may be understood as something self-sufficient or as something given, that is, as an intrinsically open structure; of course, the connotation of this statement, which bears upon the universe as a whole, is not physical but existential, as it clearly refers to the existence of the world. The latter possibility—the world as given—which is explicitly referred to in the term 'panentheism', seems to be the more appropriate starting point for a (renewed) ecological thinking. What in particular counts in this regard is whether one remains conscious of the limitations inherent in mankind's condition and, further, whether one is prepared to start deconstructing a favoured paradigm—if need be. For time and again vicious circles must be broken and fresh starts made. Thus, the beginning, the first step in constructing a paradigm, may be as important as its end, that is its final deconstruction. Moreover, as explained by Jean François Lyotard [11], the time of 'grand stories', 'meta-stories' that last forever, is over and, consequently, a theology of nature must be narrative in form, recounting history as well as legend. Evidently, a theology of nature is a constitutive part of theology. And as such it does not strive for a comprehensive intellectual panorama, a meticulously constructed philosophia perennis. Instead of presenting a panorama that pleases the intellect and is pretended to stand for ever, a theology of nature constitutes an ongoing tradition of telling old and new tales, tales that will include fragments of scientific understanding. Hence, it is a tradition that is not 'dogmatically' closed but principally open.

But for all that, the possibility of understanding the world as self-sufficient and clinging to pantheistic doctrine should not be altogether discarded. It may be useful to frame a harmonious background for ecological thinking that justifies what Arne Naess named 'quick-fix solutions', measures of a kind that can often not be avoided and that are guided by the—Heraclitean—dictum 'expect the unexpected'. Obviously, however, only pan*en*theism, as a doctrine embodying the readiness to receive the future from a realm that is out of one's reach, should at least in principle provide the proper background to criticize a Promethean arrogance about humankind's achievements and widen the spectrum of its creative resources. Thus, principally, it is open to the possible newness of things and methods, and to foster a stewardship characterized by modesty.

A case in point may be a policy that is based on the idea of 'managing' the planet Earth. For the notion of a global management need not be rejected out of hand. But a sustainable development of our planet requires strategies that can cope with surprises. Besides, instead of optimizing its exploitation, minimizing it may be a wiser choice if the integrity of the environment is to be preserved. Accordingly, global management must be 'adaptive'; it requires 'a desire and an ability to reflect continually on the values and objectives that guide our efforts', which implies (i.a.) 'the invention and implementation of technologies for sustainable development' [22]. Without this proviso, questions like 'What kind of planet do we want?' and 'What kind of planet will we make?', which may be useful by way of an orientation, will become clear indications of a presumptive attitude that has lost sight of its legitimate bounds.

Philosophizing as such does not change the world's environmental condition, but no doubt it may help us to articulate new perspectives. In order to change effectively the course of things, a duly considered environmental strategy and thoughtful tactics are needed, supported by a considerable fraction of the earth's population. But whether the time left for us is sufficient to develop effective measures is an open question. In this connection I cannot help recalling what Daniel prophetically recorded about the Babylonian king Belshazzar so long ago:

> *'They drank wine, and praised the gods of gold, and of silver, of brass, of iron, of wood, and of stone. In the same hour came forth the fingers of a man's hand, and wrote over the wall of the king's palace: and the king saw the part of the hand that wrote. ... And this was the writing that was inscribed, mene, mene, tekel, upharsin. (That is, counted [are thy days], weighed [art thou and found wanting], divided [is thy kingdom])'.*

Different versions of the text at hand were presented at the Eidos Conference 'Paradigms of philosophizing', St. Petersburg, August 1995, and at the Florensky Fund Conference 'Interaction of science, philosophy and theology in the creation of a new ecological thinking', St. Petersburg, October 1995, and also at the ESSSAT Conference 'The interplay of scientific and theological world views', Kraków, March 1996; translated into Russian this last version has been published through the care of the Barents Centre, Archangelsk.

The author owes a debt of gratitude to Dr. Iain Paul, Larbert (Scotland) for his criticism. The responsibility for the text is incumbent on its author.

Jan Fennema
Marterlaan 42
NL-1216 EX Hilversum
The Netherlands

REFERENCES

[1] **Naess, A.**, *Ecology, Community and Lifestyle* , Cambridge University Press, Cambridge 1989/1991.

[2]Clearly, there is a difference between scientific theories that, as such, can be subjected to tests, and scientific theories that, because of their all-encompassing character, function more or less as the leading ideas of research programmes. Such leading ideas will not easily be given up, and I prefer to call them meta-theories or meta-sciences.

[3] Compare, for instance, [12], [13].

[4] David Rothenberg, the translator of Naess' monograph which was originally written in Norwegian, kindly explained to me that these words genuinely spoke to him when trying to express his feelings. The words had appeared before as an epigram to a novel titled *Ararat* by the poet D.M. Thomas, written in the early 1980s, and were spontaneously remembered.

[5] The experience of the 'otherness' of things, that is the experience of the 'quality of being other' of any being distinct from one's self, is part of everyone's daily life. In particular since the philosophy of Emmanuel Levinas many books and other texts have been written on the crucial role played by the experience of otherness. In this connection reference is made only to Levinas' magnum opus *Totalité et infini* [14]. (The two key words of this title concisely call to mind the contrast between the 'totality' of the world of beings objectified by an observer and the 'infinity' of the subject of any being that opposes this observer; thus, an ontology of things loses its impact in the face of what is manifestly other.)

[6] The term 'space-time', which has been borrowed from the physical theory of relativity, is used metaphorically here. It is suggestive of a close interrelat-

edness of space and time, implying that the imminent future can be considered the 'other-in-time' in a manner consonant with the everyday conception of an arbitrary being considered as the 'other-in-space'.

[7] As innumerable texts have been written on the philosophy of the human being, only those references are given that played a special role when writing the text at hand. The topicality of philosophical anthropology was discussed in an international colloquium in 1990; a selection of the presented papers was published under the title *The Quest for Man* [15]. A study of the intrinsic meaningfulness of nature has been the subject of a thesis defended by Petran Kockelkoren entitled *De natuur van de goede verstaander* [16]. The term 'eccentric (being)', which is part of the title of a section of the text at hand, is borrowed from Helmuth Plessner's anthropology, but here used in a rather *idiosyncratic* manner; Plessner's (biological) anthropology is amply discussed in the literature mentioned above. A crucial concept of Plessner's anthropology is that the living individual—whether or not human—is characterized by its so-called intentional boundary realization; but unlike an animal, a human individual can vary this boundary, thus enabling the human species to develop culturally. A typical term used in this connection is the 'positionality' of the living individual, which distinguishes it from the non-living.

[8] The sensitivity to cherish hope and to expect determines our will and our humanity. In this connection it may be interesting to compare Jan Foudraine's criticism of those currents in psychiatry that observe (mainly) the objective/-objectifiable factors when treating the disorder of a psychiatric patient, instead of considering him/her as a human being to be met in a therapeutic community. In Foudraine's view psychiatry must loosen its ties with the natural sciences in order to heal the patient. See [17] (many reprints and several translations).

[9] The term 'positionality' has been briefly explained in note [7].

[10] Accordingly, both the human being and the world as such have been described as 'ouverture', 'overture', elsewhere; see [18], [19] (By mistake, in the last publication 'Qetiv' was printed in stead of the correct term 'Ketiv').

[11] Compare his impressive volumes [20], [21].

[12] *Merriam Webster*, Lexicon Publications Inc., New York, 1991

[13] *Woordenboek Filosofie*, Van Gorcum, Assen/Maastricht, 1992.

[14] **Levinas, E.** *Totalité et infini, Essai sur l'extériorité*, Martinus Nijhoff, Den Haag 1974.

[15] **van Nispen, J. and Tiemersma, D.** (eds.), *The Quest for Man*, Van Gorcum, Assen/Maastricht 1991.

[16] **Kockelkoren, P.**, *De natuur van de goede verstaander*, Universiteit Twente, Enschede 1992.

[17] **Foudraine, J.**, *Wie is van hout?* (Who is made of wood?), Amboboeken, Baarn 1971.

[18] **Fennema, J.**, "De l'ouverture de l'homme et du monde: réflexions

sur la technique, les sciences et la religion", *Revue des Questions Scientifiques*, **1**, 1996, pp. 27-37.
[19] **Fennema, J.**, "Ketiv und Qere. Die natürliche Theologie als Gottessuche", *Evangelische Theologie*, **2**, 1993, pp. 157-174.
[20] **Lyotard, J.-F.**, *La condition postmoderne*, Les éditions de Minuit, Paris, 1979.
[21] **Lyotard, J.-F.**, *Le différend*, Les éditions de Minuit, Paris, 1983.
[22] **Clark, W. C.**, *Scientific American*, September 1989, pp. 19-26.

PENG FU-YANG

HUMANKIND AND NATURE

INTRODUCTION

The world-wide emergence of alienation through science and technology is a direct result of humankind's lack of correct understanding of itself and of its relationship with nature. Only by realizing coordinated evolution with the natural world, by realizing that both humans and nature share a common destiny, by transforming humankind while transforming the natural world, can humans embark on a correct path through which they will eventually shake-off the many harms and crises that alienation has brought upon themselves.

ALIENATION

Alienation denotes the opposites divided from a subject while it is entering into a certain stage of development, forming an alienated external force. In the present world, almost all of them are products of scientific and technological development. Yet the advancement of modern science and technology has brought about a series of problems such as ecological imbalance, environmental pollution, land granulation, damages to the ozonosphere, greenhouse effects, energy crises, nuclear war threats, etc. These are the concomitant phenomena of science and technology which have consequently resulted in diversified and integrated crises to human existence. The present paper intends, from the stand of epistemology, to trace the origins of alienation through science and technology, probe into the coordinated evolution of both human beings and the outside natural world.

WORLD-WIDE EMERGENCE OF ALIENATION

World-wide emergence of alienation through science and technology is directly related to human ignorance of the outside world, to the lack of understanding of man himself and of the relationship between humankind and nature. Philosophically speaking, first, man has not placed himself in a right position before nature, i.e., the essentially equal status between humankind and nature was regarded from the very beginning as a subjugation between a conqueror and a conquered. Humankind is the product of nature and is in reality part of it. The early human beings, because of

© 1999 *Kluwer Academic Publishers. Printed in The Netherlands.*

the inferiority of productivity, were eventually obscure about themselves as well as about their external milieu. The absolute dependence on nature made humans unable to explain and they were in awe before such natural phenomena as thunder and lightening, floodings, earthquakes, erupting volcanoes, etc., thus inevitably deifying these natural phenomena and prostrating themselves in worshipping them. The ancient Chinese worship to those supposed supernatural beings governing heavens, mountains and waters was a vivid example. Humankind's total dependence on, passive adaptation and slavish obedience to the natural world made the relationship between them a pure one between conqueror and conquered. Yet, with the lapse of time and frequent intercourse with nature, humans gained a deeper understanding of their environment, henceforth were enabled to conquer the outer world. The application of fire was the first of such revolutions. Together with the emergence of fishery, agriculture and animal husbandry, man gained what he needed from nature through almost immense labour. This was a significant move towards humankind's recognition and transformation of the natural world, signifying that humans entered into preliminary confrontation with nature. The gradual dispelling of man's worship to and frightening before nature was eventually inevitable as his despute against nature increased.

The appearance of modern industrial civilization intensified man's contempt against nature. He began to enslave nature and waged endless and rampant plunders for wealth from surface to underground, from land to sea, from earth to space. This has inevitably caused such unexpected and irretrievable ecological crises as environmental pollution, energy exhaustion, radical decline of forest and grassland, sharp proliferation of land granulation, abnormal climate and endless natural disasters. The vindictive blows of great nature have menaced and endangered humankind's survival and development. All these are interrelated to man's misunderstanding of natural surroundings where he, being intoxicated with massive gains from and seemingly domination over nature, has totally neglected that he himself is part of it and the decisive impacts of nature and his own survival, mistaking the interwoven relationship between humankind and nature as that between a conqueror and the conquered.

The attitude of a conqueror towards nature was essentially the outcome of the evolution of man's mode of thinking from humanitarianism to human egocentricity after the Renaissance. It is quite reasonable, from the stand of humanitarianism, to expose the religious sects' usurpation of the name of God to deprive human beings of their dignity, yet the assumption of man's pursuit of worldly pleasure-seeking as the only aim of human survival would surely lead to vulgar hedonism, which, supported by the theory of human rational superiority, would again deteriorate humanitarianism into human egocentricity. Roughly there are two aspects of human egocentricity: one is the epistemological belief that theoretically there is no unsolvable problem in knowing and conquering the outside world. Historically this is true whence man is armed with modern scientific and technical achievements; the other is the practical belief that 'man is the most superior' is explained in modern terms as the affluence of man's material life and the satisfaction of man's sense organs. It is motivated by these two beliefs that man holds that the more he explored from nature the more comfortable his life would be—which would inevitably lead to usurious exploitation of natural wealth, causing rampant damages to the natural ecological pattern of existence.

Second, this defected explanation of the relationship between humans and nature has brought the relations between the subject and the object to extremity, thus narrowing man's vision of understanding. The relationship between humans and nature is in itself a two-way or interacted one: man is able to remake nature under the prerequisite of depending on and confinement by nature: nature can be remade, yet at the same time it interacts or retaliates on man. Comparatively, this interaction of nature on human beings has some special characteristics. The first is its delayed actions. In mechanics, acts and counteracts are borne simultaneously, but this is not the case as the relationship between humans and nature is concerned. For example, the massive falling of forest trees would not immediately cause water erosion and land granulation; the extensive use of chemical fertilizer would not on the spot bring about damages to human bodies. It is these superficial phenomena that concealed nature's eventual retaliation on human beings. The propaganda that 'man will surely conquer nature' is a mere manifestation of such a belief. The second is its continuance. In mechanics, forces of acts and interacts generate and vanish at almost the same time, yet the vindictiveness of nature lasts. One human who caused damage to the ecological balance could generate several, dozens or even hundreds of times of natural disasters lasting several years or even decades of years. The above-mentioned characteristics add another feature to the relationship between humankind and nature, i.e., the relationship among human beings. A triangular pattern is formed as subject (human)—object (nature)—another subject (human). The second subject may or may not be the same as the first. This pattern indicates that, when humans act upon nature, both the common interests of humans towards nature and of humans towards humans through the medium of nature would come above. In recent years, many scholars have revealed such a relationship and have pointed out that here lie the moral relations between humankind and nature. It is this pure negligence of the above two characteristics that has caused the mistaking of the two-way relationship between man and nature as mere polarization or even unilaterally of human beings against nature.

Third, the absence of correct understanding of the relations between global and regional interests and those between long-term and immediate interests resulted in one-sidedness in pursuing human interests.

Man has never ceased to pursue his own interests for his survival. The emergence of classes and states has brought about another group of interests beside those of individuals and collectives. We hold that all these should be subject to the conditions of man's milieu and global and long-term interests. Yet for a long period, humankind lacked the correct understanding of the relations between global and regional interests and those between long-term and immediate interests resulting in one-sidedness in pursuing human interests. Horizontally, humankind is pursuing individual, collective as well as state interests at the expenses of the comprehensiveness of the global interests; vertically, human beings are constantly seeking wealth from nature at the expenses of vitally important long-term interests and those of coming generations. The gradual deterioration of environment and the loss of common interests of both human beings and other creatures are the inevitable prices that human beings are forced to pay while seeking economic growth.

SCIENCE & TECHNOLOGY

Scientifically, what has become most vivid is the lack of human beings of systematic comprehension of the great nature that is a macro system having its own entirety and stratification. As the earth is concerned, the biosphere is a macro ecological system that is made up of numerous subsystems different in classes, strata and space occupation. All the units, with or without life, have their unsubstitutional functions in balance and coordination of the ecological system, each having its own significance. Any change to this existence would result in disorders of ecological functions or even chain reactions that could eventually lead to ecological crises. Henceforth, the maintenance of the balance of the entire ecological pattern is of vital importance to the life of human beings.

Yet, humankind is always unaware of the entirety and systemic character of the biosphere, and views things with or without life as isolated, static and one-sided. The way it sees not the wood for the trees has led to a series of chain damages to ecological imbalance, abnormal temperature caused by denudation and damage to vegetation, water erosion, and land granulation are vivid examples. In this world today, our arable land is consistently reducing in size while desert is consistently expanding. It would be dreadful to contemplate if man doesn't establish systematic understanding toward nature on which his life is dependent.

Second, humans do not understand that the fate of both human beings and nature is bound by common interests, a lack of understanding of the state of coexistence of both humans and nature.

Ever since they were borne in nature, humans have gained stronger and stronger ability to tame nature. Yet the state of their dependence on nature, instead of being altered, has even been testified and proved in more extensive spheres and more profound strata. The fact that humans are but a composite part of the complete system of nature determines the unity of the interests of both humans and nature. The regeneration of the life of humans is dependent on that of economy, while the latter is again dependent on that of ecology. However, in a long period of time so far, man has attached importance only to economic growth, investment of labour, capital and science and technology, and neglected the regeneration of natural resources. But as we all know, the latter is the prerequisite of the former, i.e., if the regeneration increment of natural resources lags behind the ever intensifying exploitation, man would encounter crises to the survival. Henceforth, the perfect understanding that the fate of both humankind and nature are bound by common interests, is of vital importance to the survival of human beings.

COULD ALIENATION THROUGH SCIENCE AND TECHNOLOGY BE ELIMINATED?

Basically, the division of subject from object has made possible the conquest of the latter by the former—this has been motivated by such ideas as the superiority of man's natural thinking and scientific and technical omnipotence. Yet what is also inevitable is the retaliation of the subject to the object. In the world today, it is unrealistic to eradicate such a confrontation; in other words, it would be in vain to eliminate scientific and technological alienation at a standstill. This, of course, does not imply that humankind is helpless through scientific and technical alienation. Only by coordinated evolution can humans have dialogues rather than confrontations with nature, can they gain initiatives rather than blind assaults, can they confine counterattacks from nature within their tolerance.

The concept of coordinated evolution of humankind together with nature has revealed two changes from the traditional human philosophy to ecological philosophy. First, the relations of man and nature have been changed ontologically from the state of division into the state of unity. Second, the measures of value have been changed epistemologically from man alone to the coordinated development of both mankind and nature, and at the same time the measures of ethics have been changed from the perfection of human interests alone to the perfections of both human interests and the global ecological interests.

The philosophy of the coordinated ecological evolution of both humans and nature is scientifically dealt with in four aspects: 1) nature is an organism with characteristics of diversification in species and entirety in structures; 2) with their own intrinsic objectives, any living beings perfect themselves automatically in their own ways of life; 3) there is no difference of any kind between human beings and other living beings in retaining rational operation of the entire ecological system; 4) humans are a product of nature's development and are members in the community of world living beings.

HOW TO MATERIALIZE COORDINATED EVOLUTION OF BOTH HUMANS AND NATURE?

First, according to the law of the unity of opposites, new concepts must be established that human beings and the nature do prosper and perish together, that the equality among human beings must be sublimed to the equality of both humans and nature. Human beings and nature are interdependent opposites, with one's survival and development under the conditions of the other, human beings and nature are harmonious opposites, with mutual embodiment, mutual transformation and direct assimilation. The emergence of human beings has culminated the process of the evolution of nature. Human beings can never survive without the natural world, i.e., they prosper and perish together. Better milieu of human survival and smoother economic development are preconditioned by a better ecological environment. Henceforth, the deterioration of the natural environment means nothing but devastation to the survival and development of human beings. Sociologically, human beings have never ceased to pursue human equality and freedom, or human rights as they are usually called. Yet, human rights are preconditioned by the relations between humans and nature. In case the survival milieu is endangered and hence damaged because of the deterioration of the natural environment, human rights are nothing but empty talk. Human rights are first of all the rights to survive, and further, the rights to survive are undoubtedly governed by surviving conditions. Only by subliming the equality among human beings to the height of the equality between humans and nature, and effectively coordinating the relations in between, can we humans obtain a healthy environment for the survival and development of the entire mankind.

Second, we humans should transform ourselves in the processes of transforming the outside natural world; this is logically so since we humans are but a composite part of the natural world. We have every reason to fulfill the duties and obligations in the processes of evolution. In cold winter days, we raise the temperature in our rooms in order to make us comfortable, but this cannot be achieved without consuming energy we take from the natural world. Can we then change our mode of thinking that someday in the future, with the glorious achievements of modern science and technology, we rebuild the mechanisms within our bodies to cope with the outside coldness like penguins? In this way, we can, through self-transformation of us human beings, adapt ourselves to the outside natural environment. This is what is meant by 'coordinated evolution of both humans and nature.'

Historically, the relationship between us humans and nature has gone through two stages, i.e., the stage of human worship to the natural world and that of human conquest. From the very advent of modern industry, human conquest has brought about severe retaliations from the natural world. For us human beings, it is now high time to think and examine anew the relations between ourselves and nature. Only by coordinately evolving with the natural world and ushering into a new stage of relationship in between, can we essentially extricate ourselves from the many threats and damages that scientific and technical alienation has brought on us. This too is the only correct approach to the best knowledge of us human beings.

Peng Fu-yang
Humanities Department,
Hunan University, Changsha, P.R. China
e-mail: xyx@dns.hunu.edu.cn

REFERENCES

[1] **Shixia, F.**, *Philosophical Considerations of Science Leading-fields*, Liaoning Publishing House, 1983.
[2] **Engels, F.**, *Dialektik der Natur*, 1971.
[3] **Prigogine, I. and Stengers, I.**, *Order Out of Chaos, New Dialogs Between Man and Nature*, Bantan Books Inc., 1984.
[4] **Guangming, X. and Jianming, W.**, "On the Cognition of the Origins of Environmental Problems." *Philosophical Development*, Vol. **10**, Beijing, 1994.
[5] **Goldsmith, M. and Mackay, A.L.**, *The Science of Science, Society in the Technological Age*, Souvenir Press, 1964.

FRANCIS D. O'REILLY

RURAL DEVELOPMENT KNOWLEDGE: INDIGENOUS, NECESSARY, APPROPRIATE

INTRODUCTION

This paper is concerned with the varieties of knowledge that are actually or potentially useful in the process of rural and agricultural development in the Third World. Two broad categories are recognized: the first is official, either practical or academic, external, stored in documents, testable experimentally and purporting to universal generalization; the second is informal, largely practical, largely oral, indigenous to specific times and places. In the post-colonial period throughout the Third World indigenous knowledge has rather belatedly earned official acknowledgement and respect from the world community, but there are dangers and contradictions inherent in this new status: official recognition and conservation are often a sign both that something is in danger of extinction and that it is no longer intrinsically valued by the people on the ground; official recognition also can turn indigenous knowledge into academic knowledge with all that this implies. These issues logically lead to the author's own concept of 'necessary knowledge': what type of knowledge does the contemporary peasant deem worthy of possessing and why? Decisions about what type of education children and adults in Third World villages receive raise ethical as well as economic arguments. In the collection, storage and dissemination of indigenous knowledge that indigenous knowledge is transformed into a very different category of knowledge. The epistemological implications of this process seem to be overlooked by many specialists in this important but ill-defined branch of science.

DEFINING INDIGENOUS KNOWLEDGE

Indigenous Knowledge (IK) is a concept with common but still not universal acceptance among a wide variety of disciplines and practices. Perhaps because all those familiar with the concept know what it is, there has been little felt need for any definition. A working definition might be: 'the traditional inherited knowledge of a small group of people living in a relatively small locality.' Quibbles over 'smallness' and locational specificity aside, even with this makeshift definition certain problems arise.

© 1999 *Kluwer Academic Publishers. Printed in The Netherlands.*

In the first place indigenous knowledge lacks spatial and temporal limits. Spatially IK is a phenomenon which occurs wherever people live on earth, but in practice most studies are concerned with small scale rural societies in the Third World. Temporally, IK has been characteristic of all epochs in history, yet in practice studies mainly focus on the present day, although forays in the near future into such accessible historical periods as Classical Rome or Mediaeval Europe might be predicted.

The idea of knowledge is not without difficulties. In theory everything that the peasant ('the insider') knows is peasant knowledge and, although all that knowledge may not in reality be of indigenous origin, in practice it is difficult to separate out the indigenous from the non-indigenous. Consider the following categories: knowledge of febrifugal plants; knowledge of facial scarification methods and meanings; knowledge of the types of malevolent spirits which inhabit certain trees. All three types may equally be indigenous and equally knowledge, but whereas scholars and development experts ('the outsiders') would agree that knowledge of febrifugal plants should be preserved and promoted as useful and practical, many would be indifferent or hostile to scarification knowledge and would class spirit knowledge as mere superstition. The dichotomies between abstract and concrete and between proven and believed might not have the same significance in traditional as in Western society. Bohannan noted that his organization of the religious data of the Tiv people of Nigeria reduced the diffuseness upon which the religious system rested [1]. Thus in studying IK implicit or explicit value judgements are made and areas of knowledge are selected as worthy of study and hence of conservation and promotion.

SOME QUESTIONS ABOUT INDIGENOUS KNOWLEDGE

How indigenous is indigenous knowledge? Scholars concerned with development matters are rarely historically-minded. If they have discovered a useful piece of knowledge, they are not concerned about whether that item of knowledge is one hundred or one thousand years old. Moreover, Western researchers sometimes lose their historical sense when they travel to the Third World. The assumption seems to be that the old fashioned is also the old. Two facets of Hausa (Northern Nigeria) culture, one trivial, the other of great importance, may illustrate this point. Visitors will note that Hausa men wear colourful fezzes and also that they favour wife-seclusion. Both these attributes may be accorded the reverence of tradition. Yet both date from the first half of the twentieth century when via the vehicle of colonialism educated Hausa became more aware of the current movements in the Islamic heartland of the Middle East.

This lack of historical sense can lead to great misunderstanding. Thus among the seven villages of North East Thailand where I conducted my doctoral research, one stood out as most conservative and traditional, yet it has been founded fifty years earlier by immigrants from Laos [2]. Similarly the Tiv people of Nigeria, who have been the subject of much anthropological research and are considered to have a mystical, self-defining attitude to their land, entered their homeland, Tivland, about one hundred years ago; these 'sons of the soil' have occupied their land for a shorter period than, for example, white pioneers in the American State of Oklahoma.

This historical deficiency may also lead to the labelling of everything witnessed at a given point in time as 'indigenous' or 'traditional'. The insider may be unable to assist, since he may value a strain of rice adopted by his grandfather from a colonial officer in 1920 as highly as an item of knowledge which has its origin in Pharaonic Egypt. The insider does not divide what he knows into 'indigenous' and 'non-indigenous.' The distinction may be so meaningless for him that he does not know what is new and what is ancient. Thus the Igala of Nigeria consider rice to be a traditional crop, because 'their grandmothers knew it', although its introduction in colonial days is clearly documented. Foreign imports are quickly naturalized. The *neem* tree was introduced to West Africa from India by colonialists. Now it occupies an important place in the native pharmacopoeia. Indeed Akiga, the Tiv man who wrote a valuable autobiography, noted: 'The whiteman supposes everything in Tiv culture is indigenous ... The indigenous heritage of the Tiv is comprehended in one thing and this is exchange marriage ... All the other customs ... they received from other peoples' [3]. Studying how traditional people acquired knowledge would be a task even more daunting than studying the knowledge itself. The description of the Colombian lowland pioneers provided by the novelist, Marquez, is especially insightful: 'The world was so recent that many things lacked names and in order to indicate them it was necessary to point' [4].

This historical perspective demonstrates that peasant society has been adaptive, flexible and dynamic in the past and thus if outsiders try to preserve the pristine purity of the village they may be acting contrary to native instincts. Bohannan's words in Tiv genealogy have a wider relevance: 'Genealogy has no importance; its importance is existential, that is it lies in its bearing on the particular issue being debated' [5]. Thus by extension tradition is whatever is regarded as such at any point in time. Hutheesing, studying the 25,000 strong Lisu Hill Tribe of Northern Thailand, makes the point even more strongly: 'The Lisu were continually deconstructing their own world, no definition ever held, meanings were invented or refashioned according to the whims of the moment' [6]. In such world views the existence of large areas of the unknown and unknowable is neither remarkable nor undesirable. Indeed according to Rubingh the continuity of Tiv religion 'derives so much of its strength from being surrounded by an aura of the unknown' [7]. Such vagueness and unknowingness is not a recipe for chaos or personal and social insecurity.

It is a mistake to suppose that IK is somehow incomplete and that traditional peoples are consciously deprived, just waiting for the enlightenment that supposedly higher and more accurate forms of knowledge can bring. To transpose outsider feelings onto insiders or to suppose, as John Stuart Mill did, that 'a being of higher faculty' knowing 'both sides' can accommodate the consciousness of the intellectual inferior within his expansive awareness is a gross but an old error [8]. It was articulated in the 19th century by John Davison, whom Newman called 'that grave and philosophical writer', who wrote: 'Look into the huts of savages and see, for there is nothing to listen to, the dismal blank of their stupid hours of silence; their professional avocations of war and hunting are over; and, having nothing to do, they have nothing to say' [9]. To borrow a concept from literary studies, such views may be termed 'pathetic fallacies' [10], whilst the equation of silence with stupidity denotes a peculiarly European narrowness of mind.

Of course IK scholars and practitioners generally and justifiably regard themselves as culturally sensitive and would eschew this pathetic fallacy. Though, as Turing says, 'the only way to know that a man thinks is to be that particular man' [11], they would perhaps regard "understanding the peasant" as a difficult but still attainable task through hard work and dedication; such contemporary tools as participatory observation and participatory rural appraisal would appear to be steps in the right direction. However, even when the scholar with his blistered hands sits with his peasant friends in their smoky hut after a hard day clearing bush, something is still missing: it is the keenness that hunger and want bring to perception. No matter how long the scholar lingers around the fire he always has the option to escape from the smoky hut to the world of food, comfort and opportunity. If we cannot replicate the peasant's experience in our own bodies then, in Turing's term, 'to 'signal' will have to serve ... as the only objective criterion' [12]. Two other issues on observation need to be considered: first, the idea is familiar from quantum mechanics that the act of observation may alter the observed phenomenon; second, this observed phenomenon in the village is himself an observer who incorporates the researcher into his own *Weltanschauung*—a fact which the researcher either overlooks or recoils from.

Why do we study IK? It has obviously been around for a long time yet only recently has it become an object of study. In a sense a phenomenon gains an identity and is deemed worthy of study and conservation when it is endangered. In another sense IK is part and parcel of the general world-wide environmental movement. Environmental degradation and species extinction are considered to be a result of new commercial and technical management systems coupled with demographic increase. Therefore it is assumed that prior to this widespread degradation peasant peoples throughout the world had achieved some sort of sustainable and harmonious relationship with their environments on the basis of what came to be called indigenous knowledge. Thus this IK is of great value as a means of environmental rescue. Possibly the peasant is thereby valued more for his knowledge than in his own right. Indeed this knowledge is not seen as belonging exclusively to its traditional possessor but to humanity as a whole so that it can be managed by the international community for the benefit of the world. The thrust of IK studies is therefore intensely practical rather than academic or philosophical which partly explains why certain fields of knowledge—more specifically those concerned with farming, medicinal plants and environmental management—are assiduously collected and other items of knowledge are rejected as of little import. IK is also a study in a hurry.

Apart from its flexibility and its adaptability there are also two other important characteristics of IK that need to be highlighted: specificity (time and place) and differentiation. The IK of a Thai village in 1995 is not what it was in 1975 nor what it will be in 2010—that would seem to be obvious. A river basin, such as that of the Sokoto River in North West Nigeria, albeit culturally fairly homogeneous, comprises a variety of different terrain units in terms of soil, drainage, micro-climate etc. Thus the IK of the people will vary to a greater or lesser degree depending on the micro-environment they inhabit. Broadcasting swamp rice is widespread in this river valley, yet villages several kilometres apart have a different array of rice cultivars [13]. Thus documented IK may be the result of geographical as well as value discrimination.

It is obvious that not everybody in a village is familiar with traditional medicine; there will be specialists. Less obvious is the fact that knowledge of what might seem the more basic aspects of life may also be highly differentiated. Thus in conducting a rural energy survey in Northern Nigeria [14], it soon became apparent that it was the womenfolk who were knowledgeable about types of fuelwood and their calorific properties [15]. Hutheesing among the Lisu reported how negative women were when asked questions. They invariably responded, 'I cannot say, I do not know, I am too shy to talk.' She did not see this, however, as implying a juxtaposition of power of speech in which males control and women are silenced out. Rather men and women knew the text from the inside but women were the 'prompters' and men the 'narrators' [16]. There may also be age differentiation: there are areas of knowledge that young folk are or are expected to be ignorant of.

Is IK true and accurate? We check the accuracy of items of academic knowledge we receive by cross-referencing or by experimentation. Neither of these methods seems feasible or convincing with IK. A remarkable similarity in beliefs, as for example concerning the antiseptic and tenderizing qualities of papaya leaves, may point in the direction of a botanical fact, provable by scientific experimentation, but nigh universal corroboration may simply be the result of the doctrine of signatures [17]. From the peasant's point of view, when he makes a statement about rice, he does not feel that he is making a statement about all rice but only 'this rice'—it is rice in the partitive case.

It would seem therefore that outsiders accept that, because IK exists and works, it must be true and accurate. In order to avoid epistemological controversy about the meaning of truth, truth here is taken to mean 'working truth' —beliefs about the world, which, since they seem to work without major disadvantage, can be assumed to be 'as good as true.' There is then an IK that is working truth, but is this necessarily the IK that is passed on to the outsider? The Lisu women who apparently did not know very much but, according to the researcher, had 'the text ... implanted in their bodies' have already been referred to [18]. Akiga claims: 'In former days theft was the most shameful of crimes. Telling lies, on the other hand, was always a fine thing and still is. They say, "O man, never speak the truth. If you do, you will not escape the sword. A man should be quick to answer" [19]. Peasants do not normally ask each other the type of questions researchers ask: 'How many cows do you have?', 'What diseases do they suffer from?'; they are more than a little alarmed when these questions are put to them. Nor is it easy for them to ascertain the reasons why a stranger of foreign race has entered their village and is writing down or recording their responses. In a nutshell, why should a villager pass on to a strange and incomprehensible outsider knowledge which may be personal, secret, arcane or acquired after a long apprenticeship? The general friendliness and geniality of the villagers should not delude the outsider into thinking that he is a trusted confidant. The Hausa saying *Bature yaro ne* ('the European is a boy') sums up the attitude of one ethnic group. The noun 'boy' is used advisedly. The most obvious connotations is of gullibility but there may also be ideas of boundless energy, unsatiable curiosity and lack of seriousness.

Is IK good? This may seem a strange question but it is an opportunity to introduce the idea of agricultural fundamentalism. This concept, devised by Fite in 1962, holds that there is something special and unique about the farm way of life, that farmers have a higher moral character, are more stable, dependable, honest and self-reliant, even that farming is something of a divine vocation [20]. The idea derives from the American farmer, whereas traditionally Third World farmers were neither idealized nor respected, but of late, as a result of international conservation ideals, some of this bucolic righteousness is attributed to the poor man in the village who appears to be removed from both metropolitan corruption and Western cupidity. The Sri Lankan writer, Nanayakkara, throws refreshing cold water on such ideas. In defence of the middle man and trader whom people regard as 'the very devil', she writes: 'We have come across many farmers who are not truthful in dealing, do not honour their contractual arrangements ... endeavour to pass off poor produce ... give wrong information when it benefits them ... spray ... dangerous chemicals on snake gourd, cabbage and/or beans just harvested to keep the vegetables firm and green up to the point of disposal' [21]. Doubtless the Sri Lankan rural folk do not have a monopoly on such vices.

INDIGENOUS KNOWLEDGE AS A KNOWLEDGE SYSTEM

It is all very well to deal with IK as a developmental tool and treat each individual knowledge item on its merit and in isolation, but sooner or later scholars feel the need to go beyond this and recognize and draw up systems of indigenous knowledge. Thus there are ethnobotanies, ethnotaxonomies, ethnopedologies, etc. Needless to say, the people on the ground do not spend much time discussing or refining their classification systems, but, still, it could be argued, such classification systems are implicit and none the less real for that. This may be so, but in granting that there is such a thing as an indigenous classification system, one must acknowledge at the outset that such a classification system is a very different creature from the systems which adorn textbooks.

In the first instance the basic classification of phenomena in non-European societies may be quite different from what Westerners are used to. In many of the languages of South East Asia and Meso-America nouns are classified not by gender but by a factor which relates to quantity and quality. Thus any given noun is assigned to a classifier group which seems to be derived from the shape and plastic properties of the object concerned. An example will illustrate this. The word *lem* in the Thai dictionary is defined as 'the classifier for sharp pointed objects, such as knives, axes, pins, needles, swords, etc. and also for candles, carts, books etc' [22]. Such languages often also express abstract ideas by concrete examples.

If we examine any of the major scientific systems of soil classification, we will see, in Niemeijer's words, that 'they tend to focus on the deeper soil horizons' and 'they strive for universal applicability and often make use of pedogenetic criteria for their semantic organization' [23]. Since they build up complex patterns through the accretion of simpler building blocks, there are empty cells in the matrix which can be filled by soils perhaps not yet met but imagined. Any new soil can be placed in the appropriate place in the matrix. Traditional ethnopedology makes no such claims to universality. In the first place the idea of the soil as the medium in which plants grow may not be clearly distinguishable from other site characteristics. Thus Lynham *et al.* complain that Mauritanian farmers frequently make use of the same names for soil and landscapes [24]. In the evaluation of soil proper, often identified by surface horizons and superficial characteristics, the traditional farmer is concerned with land use potential. As well as being site- and application-specific, any resulting indigenous soil maps are dynamic in contrast to the static scientific soil maps [25].

The situation with regard to plants and animals is even more complex. Scientific taxonomy is hierarchical, based on homology rather than analogy. It uses 'basic anatomical design', not ecological classification as the criterion of higher-level classification [26]. Nobody should doubt or belittle the botanical knowledge of the peasant. Stross claims that Mayan five year olds are able to classify over one hundred 'botanical types' [27]. However, Olson sees these elaborate indigenous taxonomies as 'based on morphological aspect and ecological function rather than reproductive lines of descent.' Moreover, they 'make no attempt to relate such classes to a unified conception of Nature', which he considers 'the major concern of science since Aristotle' [28]. The concept of 'a weed' for example, explicable in many Afro-Asian languages only by means of tortuous neologisms, certainly has a quite different significance in rural societies where the voluntary vegetation (and the pests) may form part of an augmented or alternative harvest [29] [30]. Knowledge of plants and animals in peasant villages is not, however, solely based on ecology or shape; consideration of totem, taboo and (to outsiders) bizarre kinships play a part and are not to be treated as negligible curiosities. Clearly therefore indigenous taxonomies will be very geographically specific, shaped not only by peculiar environmental conditions but also by such matters as history, religion and myths. It would be understandable for the hard pressed development researcher to postpone mastering the taxonomic complexity of his study area and to concentrate instead on the application of scientific knowledge to the 'problems.' This would be a mistake. First of all, Western scientific semantics are meaningless to the peasant especially since they do not seem to relate to what can be perceived directly by the senses, and the large scale scientific education of the peasantry is neither possible nor worthwhile. On the other hand, as Gonzalez modestly puts it, 'the combination of different types of information—whereby the scientists' wording and the farmers' practical experience are integrated ... is helpful in defining appropriate development interventions' [31].

The fact that traditional classification systems are concerned with the here and now does not mean that they cannot come to terms with new ideas, new crops and animals. It has been remarked that strangers who wish to settle in an African village are usually adopted by one of the families. The stranger is absorbed and his danger neutralized by 'making him one of us.' This idea of kinship can be extended to indicate the way in which IK colonizes the unknown. An example will make this clear. The Thai word for 'rice' is *khao*. The dictionary also lists such words as *khao fang* (millet) and *khao salli* (wheat) [32]. What the Thai have done is to identify foreign cereals as types of rice; the alien plants have been neutralized; they have become one of us. Thus foreign phenomena are incorporated into the body of IK not by studying the intrinsic qualities of these phenomena but by seeking out similarities with existing known phenomena. Although Thai and Hausa have no possible linguistic kinship, this activity is characteristic of both. Both make extensive use of the universal solvent, water, as a universal prefix; both use part of the body or body functions to describe the greater world outside the human body [33] [34].

Indigenous awareness of phenomena may also bespeak a different way of knowing. The fresh groundwater lenses of Pacific atolls such as Kiribati are a remarkable hydrological phenomenon. Such systems 'have a zone of fresh groundwater ... overlying seawater with a transition zone which may vary from less than a metre to over hundred metres in thickness depending on recharge rate, porosity and the lateral dispersion coefficient' [35]. The lens is called the Ghyben-Herzberg lens and the relationship of variables in the system is understood by the Ghyben-Herzberg relation or approximation which relates the height of fresh water to densities of fresh and sea water [36] [37]. As well as the scientific interest such lenses are of great practical importance, since they are often the only source of fresh water on these islands. Probably most I-Kiribati would have as much difficulty in understanding the Ghyben-Herzberg formula as they have in pronouncing that hyphenated Germanic name. Yet clearly in their own way they were aware of this (largely unseen) phenomenon and of its basic mechanics; without such an awareness their civilization, nay their very lives, would have been impossible. The term 'heuristic' (from Greek *Eureka*) has been coined to describe such ways of knowing [38].

NECESSARY KNOWLEDGE

It is clear that IK is almost everywhere in retreat. Another example from Kiribati will illustrate this. The I-Kiribati had developed a massive navigational lore which included such skills as being able to identify land by studying the colour of cloud bases. With powered boats and mechanical navigational aids most of these skills have been lost so that these days young men who set out to the sea in canoes are liable to get lost (sometimes irrevocably) as soon as they are out of sight of the shore [39]. Such examples could be multiplied by case studies from Africa, Asia and Latin America. It would be easy but unjust to condemn such peoples for their cultural negligence. The concept of necessary knowledge may help.

Necessary knowledge (the present author's neologism) is knowledge that is necessary for survival in the broader as well as in the life or death sense. For most of the history of traditional peoples environmental knowledge of the most intimate kind was necessary for survival (in both senses). These days a knowledge of bureaucratic, administrative or mechanical environments may be more essential to the farmer than the identification of woody plants.

Examples from Western society may increase our tolerance for the amnesiac peasant. There is an area of human knowledge which we have all experienced and largely forgotten: the field of children's lore and games. In making this analogy I am not of course suggesting that peasants are childlike but that 'conceptual development is orderly whether that development proceeds historically or ontogenetically' [40]. Children's lore is truly IK in that in the main it does not come from books and is rarely written down (except by a few researchers [41]). It is passed down from elder to younger siblings perhaps with gender differentiation and not without contextual modification. Attempts by such as teachers or other outsiders to appropriate or share in this knowledge are doomed to failure; the teacher is left clutching risible fossils whilst secret knowledge evolves into new forms, foreign and self-protective. However, as soon as the child reaches adolescence, the whole corpus of knowledge is rejected, except perhaps by those who are considered to be immature, maladjusted or 'strange.' This voyage into childhood brings out two germane points: how easy it is to forget and even despise knowledge when we grow out of it or when we adopt what seems to be a more appropriate way of looking at the world; how the appropriation of 'our knowledge' by authority figures leads to a serious devaluation of that knowledge.

An adult and more specific example from my home county of Lancashire may be a better illustration. When I was young, the main fuel in the homes was coal. The Lancastrian housewife had an encyclopedic knowledge of the coals on the market and their calorific qualities (not of course coal in general nor coal in any scientific sense). Now the vast majority of homes use gas or electricity and the majority of young housewives would not know how to light a coal fire let alone recognize different types of coal (or in some cases coal at all). I have a friend in London, a retired teacher, who is interested in coal, has books about it and discourses at length on the different types. He does not burn coal; so for him coal knowledge is purely an antiquarian, academic or hobby interest. Unless one is a very impassioned hobbyist, it is very difficult to sustain and maintain a high level of useless knowledge but useful knowledge can be absorbed and internalized in vast quantities without even trying. Academics may be resistant to this fact, if fact it be, precisely because much of their lives is taken up with the learning and teaching of knowledge which does not seem immediately useful.

Acceptance of the idea of necessary knowledge may help in coming to terms with the paradox implicit in the question: does the Third World farmer have a good or a bad memory? In certain cases, such as in reciting genealogies, his feats of memory are prodigious (there may be some inverse relationship between memory power and literacy [42] [43], though at times elaborate mnemonic devices are used [44]). In other instances he appears to be a creature of almost unimaginable forgetfulness. I have heard the response 'don't know' given to the following questions among others: 'How old are you?', 'How many children do you have?', 'What size is your farm?'. Of course, in certain cases the farmer might be politely saying, 'It's none of your business,' but in general there does seem to be a lack of precision with regard to mensuration and methodologies. This lack of precision does not, however, mean a lack of knowledge. It is possible to perform tasks successfully without being able to explain how success was achieved or to know a great deal about a given topic without knowing that one knows it. Such type of memory, dubbed 'largely procedural and tacit' by psychologists [45] [46] [47], applies not merely to the homely examples of touch typing or tying Windsor knots or to the impressively varied Nigerian cuisine where 'recipes as such are not part of traditional cookery techniques' [48], but also, perhaps surprisingly, to the world of the human stock market dealers and their computer simulation models [49]. Could the use of mastery of grammar (deemed innate by some psychologists [50]) by unschooled native speakers be an even grander example of this? Learning in such contexts may be partly a case of 'watch what I do and try to do it yourself' and partly developing instincts and reflexes through prolonged immersion in a body of data or repeated performance of tasks, not necessarily inherently simple but capable of being broken down into simple constituent parts—perhaps the claim by the steelworkers of Sheffield (Northern England) that they could 'smell' the quality of steel is thus explicable. With such a wide and sophisticated applicability of tacit memory it does not seem farfetched to propose it as at least a partial explanation of why the peasant who can do so much can explain so little. Acceptance of the idea of tacit memory also, incidentally, exposes the schoolmasterly diktat 'if you can't explain it, you don't understand it' as a canard, both illogical and arrogant.

THE TRANSFORMATION OF INDIGENOUS KNOWLEDGE

A number of possible fates awaits IK. It may die out and, since the world is so large, change so rapid and time so short, that is likely to be the plight of much of it. Some, however, will survive on its own terms; it is remarkable, by the way, that types of English country IK still survive among modern Londoners, e.g., plant antidotes for nettle rash. Third World IK has to date spawned few native antiquarians or academic saviours, but an increasing amount of such knowledge is being collected by researchers and stored in enormous databases, whence it is disseminated by means of electronic or periodical networks. It may even thereby be reintroduced into areas where it had become extinct.

A practical distinction is made between knowledge about an object, which is stored in IK databases, and the object itself, which, in the case of indigenous cultivars, is stored in international genebanks. This practical, administrative, political and professional (genebanks for geneticists, knowledge banks for social scientists) distinction should not blind us to the fact that 'the thing' and 'knowledge of the thing' are intimately connected in village reality. Therefore experts of the knowledge bank and the genebank should keep a close eye on one another. In this context the genebank ethic of keeping genetic material 'in trust' for the Third World farmer and the genebank realization both that landraces can be the raw material for scientific advance and the saviours of genetically overspecialized commercial crops may have lessons for the knowledge bank [51]. A further point, the patenting of life forms made possible by the Uruguay round of GATT, may have profound consequences for both genebanks and knowledge banks, especially if IK comes to be regarded as intellectual property [52]. The effects of such patenting on peasant agriculture are likely to be considerable.

Two questions need to be asked about collected IK: Does the international body of knowledge, removed from, if not neglected by, its progenitors, deserve to be called indigenous knowledge in any meaningful sense of the words or should a new term 'knowledge of indigenous origin' be coined? Is it worthwhile making any distinction within the body of collected IK between 'knowledge of what the peasants know' (knowledge about knowledge) and knowledge of indigenous origin which is absorbed in its own right as an augmentation of Western scientific knowledge? To the two concepts, original IK and transformed IK, should be added a third, noted but not pursued: IK as an academic discourse with its own *ordre de discourse* [53] 'that has its borders set not unlike the Roman empire' [54].

IK scholars are well served by a variety of centres: the 'Indigenous Knowledge and Development Monitor' (a specialist periodical) lists twenty-two of these (all contactable by fax, some by e-mail) and thus neither the researcher testing his pet theory nor the student writing a term essay need be short of easily acquired data. More important, however, is the dissemination of knowledge to the farmers. A number of methods may be mentioned, all deserving of lengthy and serious evaluation: the use of 'strange farmers' as part of a North-South or South-South hands-on technology transfer [55]; the innovative Canadian Farm Radio which disseminates IK development scripts to radio stations throughout the world [56]; the compilation or updating of indigenous knowledge systems (such as botanies) or indeed indigenous encyclopaedias [57] [58] [59]; the pioneering of multi-media technology (especially useful in inaccessible rural locations) [60] and knowledge-based systems, computer programs which contain both declarative and procedural knowledge [61]. The interface between IK and national extension services and the contribution of IK to national agricultural policies are particularly important as well as highly sensitive. In addition to the aforementioned methodologies, which have as their targets rural men and women, attention should also be given to the role that IK can play in education, both general and vocational, at school level [62], and in the area of higher education [63] [64].

ENDPIECE

In writing this piece I have like the latter day Kiribati fishermen tried to stay 'in sight of the shore.' I have been ever conscious that every point made concealed entangling depths and that case study material, of necessity selective, could easily elicit the response: 'not in my village!' In a sense this is symptomatic of the state of IK studies—a wide array of 'just so stories.' Sciences proceed through recognizable stages, such as data collection, description, analysis and evaluation, generalization and 'law making', perhaps finally 'law breaking' and morbid introspection or even the unenviable state of affairs that Glanville attributes to cybernetics, where outsiders rifle the academic 'toolkit' and 'having lost the coherence as well as the contents of our toolkit we are left with nothing' [65]. Arguably, IK studies are at the stage of data collection and tentative description. However, collecting IK items is not like collecting beautiful shells on the beach and carefully cataloguing them for detailed study some rainy day. Elusive, mutable, fragile, corruptible, IK demands a new scientific methodology. In the midst of their databases it is incumbent upon researchers to face the blunt question: What are we up to?

Dr. F.D. O'Reilly
London Guildhall University and
Science Technology Development Forum, Glasgow

REFERENCES

[1] **Bohannan, P.**, *Peoples of Africa*, New York, 1965, p. 59.

[2] **O'Reilly, F.D.**, *The process of innovation in the subsistence agriculture of north-east Thailand with particular reference to the Lam Pao Irrigation Area, changwat Kalasin*, Ph.D. University of London (unpublished), 1974.

[3] **Akhirghirga, S.**, *The history of the Tiv people*, (unpublished MS, no date), translated and cited by Rubingh, E., Sons of Tiv: a study of the rise of the church among the Tiv of Central Nigeria, Baker Book House, Michigan, 1969, p. 141.

[4] **Marquez, G.G.**, *One hundred years of solitude*, (translated from Spanish by Rabassa, G.), Picador, 1979, p. 9.

[5] **Bohannan, P.**, "A genealogical charter", *Africa*, **Vol. XXII**, Oct. 1952, pp. 301-315.

[6] **Hutheesing, O.K.**, "Male jargon, female talk: verbal exchanges of academe and the mundane", *Kajian Malaysia*, **Vol. XII**, Nos. 1 & 2, June/Dec., 1994, pp. 185-209.

[7] **Rubingh, E.**, see [3].

[8] **Mills, J.S.**, *Utilitarianism, on liberty, considerations on representative government*, ed. Williams, G., The Everyman Library, London, 1993, pp. 8-12.

[9] **Newman, J.H.**, *The idea of a university*, Oxford at the Clarendon Press (based on the 9th edition of 1889), 1976, pp. 149-150.

[10] **Umar, H.D.**, *Man and Nature in Hardy's fiction*, B.A. dissertation (BUK, Nigeria), 1977, unpublished.

[11] **Turing, A.M.**, "Computing machinery and intelligence", *Mind*, **59**, pp. 433-60, cited in **Michie, D.**, Consciousness as an engineering issue, Part 1, Journal of Consciousness Studies, **1**, No. 2, Winter 1994, pp. 182-95.

[12] **Turing, A.M.**, *op. cit.*

[13] **O'Reilly, F.D.**, field observation, 1981.

[14] **Cline-Cole, R.A.** *et al.*, *Wood fuel in Kano*, UNU Press, 1990.

[15] **O'Reilly, F.D.**, field observation, 1982.

[16] **Hutheesing, O.K.**, *op. cit*

[17] **Schefold, R.**, "Function follows form: on the ritual efficiency of plants and textiles in Indonesia", *Indonesia Circle*, **No. 66**, 1995, pp. 79-99.

[18] **Hutheesing, O.K.**, *op. cit.*

[19] **East, R.** (translator and editor), *Akiga's Story: the Tiv people as seen by one of its members*, OUP, 1965, p. 344.

[20] **Fite, G.**, "The historical development of Agricultural Fundamentalism in the 19th century", *Journal of Farm Economics*, **44**, 1962, pp. 1203-11.

[21] **Nanayakkara, U.**, "Difficulties in entering into a market economy in Sri Lanka with special reference to agricultural produce marketing", *Marga*, **Vol. 13**, No. 1, 1994, pp. 45-85.

[22] **McFarland, G.B.**, *Thai-English Dictionary*, Stanford University Press, Stanford, 1969.

[23] **Niemeijer, D.**, "Indigenous soil classifications: complications and considerations", *Indigenous Knowledge and Development Monitor*, Issue **1**, April 1995, pp. 20-21.

[24] **Lynham, M.B.**, *et al., Farming systems research along the Senegal River Valley: a rainy season reconnaissance survey in Guidamaka, Gorgol and Brakna regions*, University of Arizona College of Agriculture, cited in **Niemeijer, D.**, *op. cit*

[25] **Niemeijer, D.** *op. cit*

[26] **Gould, S.J.**, *Wonderful life: the Burgess Shale and the nature of history*, Penguin Books, 1989, pp. 9, 8-99 & pp. 209-214.

[27] **Stross, B.**, *Speaking of speaking: Tenejapa Tzeltal metalinguistics*, in R. Bauman and J. Scherzer (eds) Explorations in the ethnography of speaking, CUP, 1974, cited in Olson, D.R., The world on paper, CUP, 1994, p. 221.

[28] **Olson, D.R.**, *op. cit*

[29] **Irens, G.W., Moody, K., & Egunjobi, J.K.**, *West African weeds*, OUP, 1978.

[30] **O'Reilly, F.D.**, "Alternative Harvest", *Maapallo*, **No. 4**, 1988, pp. 13-15.

[31] **Gonzalez, R.M.**, "KBS, GIS and documenting indigenous knowledge", *Indigenous Knowledge and Development Monitor*, **Vol. 3**, Issue 1, April 1995, pp. 6-7.

[32] **Haas, M.**, *Thai-English Student's Dictionary*, Stanford University Press, Stanford, 1967.

[33] **Haas, M.**, *op. cit*

[34] **Abraham, R.C.**, *Dictionary of the Hausa language*, Hodder & Stoughton, London, 1978.

[35] **Chapman, T.**, "Hydrological systems and processes", in: **Falkenmark, M. & Chapman, T.** (eds), *Comparative Hydrology: an ecological approach to land and water resources*, UNESCO, 1989, pp. 60-61.

[36] **Wright, E.**, "Small flat islands", Ch. 18 in: **Falkenmark, M. & Chapman, T.**, *op. cit.*

[37] **Barker, J.A.**, "Freshwater-saltwater relations", *Proc. Regional Workshop on Water Resources of Small Islands, Suva, Fiji, Commonwealth Science Council Tech. Publ.*, No. **154**, pp. 124-130.

[38] **Pawluk, J.** *et al.*, "The role of indigenous knowledge in agricultural development", *Journal of Soil and Water Conservation*, **4**, 47, pp. 298-302.

[39] **Evans, T.**, Taborio, Kiribati, personal communication, 1994.

[40] **Olson, D.R.**, *op. cit.*, p. 122.

[41] **Opie, I. & P.**, *The lore and language of schoolchildren*, Oxford, 1959.

[42] **Havelock, E.**, *The literate revolution in Greece and its cultural consequences*, Princeton University Press, New Jersey, 1982.

[43] **Carruthers, M.J.**, *The book of memory: a study of memory in mediaeval culture*, CUP, 1990.

[44] **Kiebel, C.B.**, "Memory sticks and other mnemonic devices", *The Nigerian Field*, **Vol. 55**, Parts **3-4**, 1990, pp. 91-98.

[45] **Michie, D.**, "Consciousness as an engineering issue", Part 2, *Journal of Consciousness Studies*, **2**, No. 1, 1995, pp. 52-66.

[46] **Squire, L.R.**, *Memory and brain*, OUP, 1987.

[47] **Anderson, J.R.**, *Cognitive psychology and its implications*, 3rd ed., New York, W.H. Freeman, 1990.

[48] **Anthonio, H.O. & Isoun, M.**, *Nigerian Cookbook*, Macmillan, 1982, p. 11.

[49] **Kelly, K.**, *Out of control: the new biology of machines*, Fourth Estate (London), Ch. **22**, pp. 420-449.

[50] **Pinker, S.**, *The language instinct: the new science of language and mind*, Penguin Press, 1994.

[51] **O'Reilly, F.D.**, "Stolen rice", *Science, Technology and Development*, **Vol. 13**, No. 1, April 1995, pp. 114-115.

[52] **Jain, H.K.**, "GATT proposals and India's seed policy", *Seed Research*, **Vol. 20**, No. 2, 1992, pp. 65-69.

[53] **Foucault, M.**, *The archaeology of knowledge* (trans. from French by Smith, A.M.S.), Tavistock publications, 1979.

[54] **Vranckx, A.**, personal communication, cited in: **O'Reilly, F.D.**, "More on 'Why conferences?", *Science, Technology and Development*, **Vol. 11**, No. 3, Dec. 1993, pp. 353-355.

[55] **Benzabih, H.A.S.**, *Environmental and socio-economic constraints of rain-fed agricultural land settlement projects in marginal climatic zones: a case study of the Jabel al-Akhkdar Upland, Libya*, Ph.D. University of London (unpublished), 1987.

[56] **O'Reilly, F.D.**, "Farm Radio Network", *Science, Technology and Development, op. cit.*

[57] **Dalziel, J.M.**, *A Hausa botanical vocabulary*, London, T. Fisher Unwin, 1916.

[58] **Berthelsen, C., Mortensen, I.H. & Mortensen, E.**, *Kalaalit Nunaat Greenland Atlas* (English translation Jones, W.G.), Pilersuiffik, 1990.

[59] **East, R. & Imam, A.**, *Ikon Allah: Labarin halitta iri iri ta cikin duniya* (The power of God: the book of created things in the world), Zaria, Nigeria, 1952.

[60] **Wratten, R.**, "Cluster exchange / information well / transfer-box applications", *International Journal of Rural Studies* (IJRS), **Vol. 2**, No. 1, April 1995, pp. 9-11.

[61] **Gonzalez, R.M.**, *op. cit*

[62] **O'Reilly, F.D.**, "Rural children's education and the Green Revolution in Nigeria", *Journal of General Studies*, **Vol. 3**, No. 1, 1982, pp. 193-199.

[63] **Singh, M.** (editorial), "Proposal for autonomous rural degree college: practical considerations", *IJRS*, *op. cit.*

[64] **O'Reilly, F.D.**, "Towards a course on research techniques for rural development", *IJRS*, *op. cit.*

[65] **Glanville, R.**, "A (cybernetic) musing", *Cybernetics and Human Knowledge*, **Vol. 2**, No. 4, 1994, pp. 53-55.

H. THOMAS R. MARK

THE PREDICAMENT OF ECOLOGICAL-ECONOMIC VALUATION AND THE NEED FOR LINKAGE THROUGH AN INSTITUTIONAL FRAMEWORK

INTRODUCTION

Human societies have evolved from small scale subsistence to the present state of globally interdependent societies. This development has transformed not only societies in terms of their scale but also ecological systems (ecosystems) which have been substituted by human induced environments in order to provide natural resources and space to societies. This ancient drive to survive and improve the well-being of humans is increasingly becoming ecologically unsustainable. Economic development is being made at the expense of the health of ecosystems, reflected in the reduction of biodiversity, resilience, etc.

This predicament is here traced to the absence of understanding the connection between the creation of economic value and organization and to the connection between the economic and ecological systems from an ecosystemic perspective. I suggest an interpretation of how organization and institutions emerge in the economic system by using the ecosystem as a model. Both systems are seen as complex and evolving hierarchies of processes for the purpose of understanding the role of monetary institutions in promoting a development that is ecological and economical non-sustainable. The nested hierarchical organisation of the spatiotemporal scales are similar in both.

Temporal scales arise when economic agents act under uncertainty. These scales become hierarchically organized as agents which maximize economic gains per unit of time and extension in time through entrepreneurial and management activities. Temporal scales are always associated with spatial confinements.

Furthermore, I suggest that monetary institutions provide the range of spatiotemporal scales in which the economic system develops. The introduction of monetary institutions on a national level in Britain during the 17th century and on an international level in the early part of the 19th century appears to have facilitated economic growth within the range of spatiotemporal scales provided by monetary institutions as an incentive structure.

© 1999 *Kluwer Academic Publishers. Printed in The Netherlands.*

Industrial societies have been able to increase their spatial scales without increasing the temporal scales emerging from their monetary institutions. This has led to a deviation between the correlation of spatiotemporal scales in which societies operate and the correlation of ecosystem scales. Industrial societies have become nested economic systems instead of nested ecological-economic systems. Monetary institutions need to be designed to promote economic interaction with the ecosystem so that they emulate the spatiotemporal scales of the ecosystem.

A REFLECTION ON THE HISTORY OF ECONOMIC THINKING

Adam Smith, the founding father of economics, wrote more than two hundred years ago *'The things which have the greatest value in use have frequently little or no value in exchange . . . Nothing is more useful than water'* [9]. In a way, Smith pointed at the failure of the economic systems to reflect the values of more fundamental realities such as natural systems (ecosystems). Classical economists explored values not only as the result of supply and demand but also in terms of an underlying supply structure.

Later, the neoclassical economists solved the discrepancy between *value-in-use* and *value-in-exchange*. They introduced the idea that all economic values are the result of marginal adjustments of shifting demand and supply quantities in the market approaching an equilibrium [9]. Economists thereby abandoned the idea that economic value could or even should reflect a structure beyond demand and supply. Attempts to link value and structure did not, however, completely cease with the introduction of neoclassical theory:

- In the late 19th century, Böhm-Bawerk, suggested a link between economic structure and time [9]. He concluded that increasing economic complexity, which he called roundaboutness, increased the time to produce products, which also increased the output-economic value.
- In the first decades of this century, Joseph Schumpeter [19] described the wealth creating process as a dynamic evolutionary process of cycles of birth, death and renewal of structures. He emphasised that structures are created and destroyed in the process of creating economic values.
- In 1932, Ronald Coase suggested that there is a cost of transacting involved in bringing people together in order to exchange goods or services—values—with each other [5] [6]. He suggested that this

invisible cost explains the existence of organizations such as companies, which can be more cost effective instruments than markets. Coase thereby linked the value of exchange to the structure of organization.

THE ECOSYSTEM AS A MODEL

The human induced organizational mechanism that has been instrumental in the evolution and transformation of society and the earth system may have similarities with the organization of ecosystems as nested hierarchies. Models that describe ecosystems as complex evolving hierarchies have been suggested by some ecologists. I will extend the idea that ecosystems are nested hierarchies in order to understand the structure and dynamics of the economic system.

The ecologist C. S. Holling [10] has interpreted the successional change in ecosystems from the same perspective as Schumpeter [19]. Both Schumpeter and Holling emphasize the dynamic interaction between micro and macro levels. Holling's evolutionary model [11] builds on an understanding of ecosystems as organizations consisting of small and fast constituents at the micro level and large and slow constituents at macro level. Time plays a role through the different process rates—*temporal scales* —at the micro and macro level and through the dynamic progression of the development cycles. This model explains how system complexity evolves from the cycles of birth, death and renewal, by which the fast micro levels are first mobilized to be followed by slower macro levels.

Holling's model complements the predominating stability model of ecosystems [16]. This model emphasizes the stability aspect of system complexity. Stability is achieved by higher or macro levels correcting for errors or inefficiencies at the level below operating at a faster process rate or shorter temporal scale.

These two models provide a comprehensive understanding of ecosystems as both evolutionary and stable hierarchical systems.

LEARNING ABOUT UNCERTAINTY

In this paper I emphasize that economic structures emerge from time delays—temporal scales—reflecting the cognitive ability of humans to learn about and from economic opportunities.

Economic agents, whether they are individuals or companies, seek economic opportunities for the purpose of increasing their economic gains. This is done under some degree of uncertainty reflecting the diversity and complexity of the economic system.

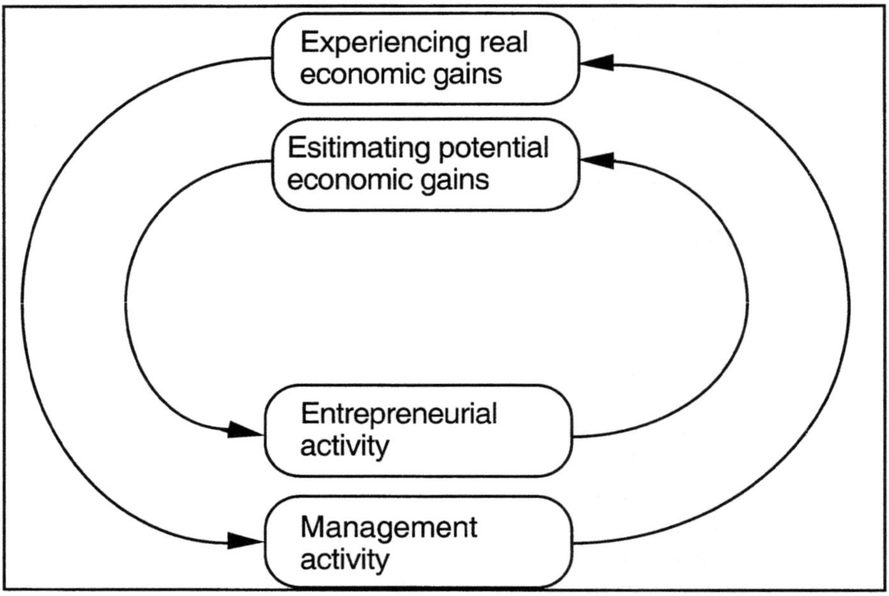

Figure 1. Economic opportunities as a feedback process, introducing temporal scales

Successful agents learn, however, to reduce their degree of uncertainty by absorbing the experiences of each opportunity for the benefit of the next one. Figure 1 illustrates this process of learning (compare with process rates in ecosystems), in which each agent's effort is fed back with a time delay, as a gain or loss. Sorting economic activities into entrepreneurial and management opportunities is not new, but I suggest that they should be interpreted as feedback processes with time delays or temporal scales in order to understand how economic structures emerge.

Entrepreneurial opportunities are important because every opportunity cannot be maintained indefinitely in a dynamic and evolving economic system. Any agent that wishes to maintain gains needs to risk them by investing them in entrepreneurial opportunities that can supersede declining management opportunities.

Management opportunities are important because the objective of each economic agent to maximize returns per unit of time and extension over time—that is to make as much money for as long as possible. This objective is not likely to be achieved by pursuing just one opportunity, since agents are commonly presented with a choice of opportunities of short duration and high returns or long duration and low returns. Instead, agents need to establish many opportunities of varying earning power and duration, i.e., temporal scales, simultaneously.

Economic systems should therefore be understood as both evolutionarily generated by economic agents' entrepreneurial function and hierarchically generated by their management function.

THE NEED FOR ORGANIZATION AND INSTITUTIONS

The need for organization emerges in order to maintain a structure in which the opportunities will support and complement each other. Assume that there are two opportunities in an economic system, one more uncertain opportunity **A** provides a higher return per time and one safer opportunity **B** provides a lower return over a longer time for the economic agent. **B** makes it possible to extend **A** in time and **A** makes it possible to increase the return of **B**.

This mutual reinforcement makes it beneficial for economic organizations to establish hierarchies of entrepreneurial and management opportunities of varying temporal scales. Economic complexity evolves from an entrepreneurial function, that extends the life span of an economic agent, and a management function, that maintains stability through control.

Ronald Coase has realized that there is a cost of transacting and that this cost can be reduced through the existence of rules in organizations. Coase's findings led to a renewal of Institutional economics, pointing to the role of society's institutions (rules) in promoting social and economic development at large. North [14] [15] emphasized an evolutionary perspective of the formation of organizations and institutions, that it is the incentives embedded in the institutional framework that direct the process of learning by doing, which changes the organization as well as the institutions of the economic system.

An interpretation of the organization of the economic system as a nested structure of temporal scales suggests the following. Each organization exists in a matrix of other organizations, in which each organization's range of temporal scales is adjusted to the range of the surrounding organizations. It is the institutions of society that set the range of temporal scales that the economic system at large operates within.

HOW ECONOMIC STRUCTURES EMERGED IN THE 19TH CENTURY

Nathan Rosenberg [17] [18] has described from a technological perspective how economic structures emerged in the industrializing countries of the 19th century. *'The growing productivity of industrial economies is the complex outcome of large numbers of interlocking mutually reinforcing technologies, the individual components of which are of very limited economic consequences by themselves'* [18].

Rosenberg outlines technological change as a process of accumulation of small complementary improvements that mutually reinforce each other. The total benefit of one innovation has to await the achievements of other innovations. For one technology **A**, the benefit of cost reductions will increase with the possibility of interlocking it with technologies **B**, **C**, etc. Rosenberg exemplifies this with the payoffs of the innovation of the railway, which were partly derived from the increased productivity in agriculture due to increased scale and regional specialisation. The economic historian J. R. Beniger [1], has also described how the economy of scale in America, started to increase with the introduction of the railway in the 1840s. The scale was sequentially spread to other sectors of the economy. Thus the reduction of transportation costs expanded the market and increased the opportunities for scale economy.

Interlocking more and more technologies with each other, i.e., creating larger and larger clusters of technologies, increases the range of temporal scales as well as the range of spatial scales of the economy. This is because changes in one part of the system will affect other parts with increasing time delays as the number of interlocking technologies increases. That is, change in technology **A** is fed back to itself relatively quickly, whereas the feedback from **B** to **A** will take longer.

These observations on interlocking technologies support the idea that the economic system is a nested hierarchical system of spatiotemporal scales, whose complexity grows with its increasing range of scales.

THE ROLE AND EVOLUTION OF MONETARY INSTITUTIONS

The institution of money has not only passively adjusted to the increasing complexity of society, it has also promoted it and sped up the transformation of economic systems. The institution of money reduces uncertainty and thereby facilitates exchange over both space and time. The introduction of monetary institutions operating on an increased range of scales appears to have facilitated economic development within that range by acting as an incentive structure.

I will here briefly explore the historical influence of monetary institutions on the spatiotemporal scales of society by using the example of the British economy as a forerunner of global economic and ecological transformation.

Access to energy has played an important role in economic development [22]. The industrial revolution took off with the invention of a technology that enabled the use of the energy in coal, but its impetus had been building up over a rather long time. England already displayed signs of industrial development more than a century before the industrial revolution of the late 18th century and the transition to coal energy, in metal works, paper mills, mining, etc. Economic historians trace the roots of the transition from an agrarian to an industrial society to the commercialization of agriculture that began in England with the *first enclosure movement* of agricultural land (privatizing the commons) in the late Middle Ages. The agrarian structure was thus influenced by the monetary economy long before the industrial revolution [21].

National monetary systems centered around national banks were first developed in the late 17th century. The founding of the Bank of England in 1694 added the function of issuing money to the previous functions of banks of merely clearing and holding money. This increased the quantity of standardized money in the British economy which led to the establishment of local banks all over Britain in the 18th century and an increased availability of money [0]. Braudel [4] concluded that the Bank of England became a wheel around which the whole British economic system moved. This stimulated growth of the domestic economy in the 18th century, facilitating the transformation from an agrarian to the first industrial society.

In the 1820s, the Bank of England established an informal gold standard. Money markets for international trading and investment banking to finance large scale railroad projects etc. were established in London during the same time [8]. Thus the British monetary economy had started to expand outside the geographic borders of the country. As this expansion picked up speed in the middle part of the 19th century (Table 1) with larger and larger areas embraced by monetary standards and instruments, a global economy began to emerge that became an integrator of societies. Likewise, the temporal scale increased with the establishment of new credit systems, such as investment banking. At the same time the introduction of monetary instruments, such as money markets, increased the access and speed of monetary transactions, which meant that the ability to carry out monetary transaction over a wider range of spatiotemporal scales increased.

TABLE I

United Kingdom's relative share of world manufacturing output [13].

Year	1750	1800	1830	1860	1880	1900
%	1.9	4.3	9.5	19.9	22.9	18.9

Later developments have mainly brought further increase of the speed to the economic system. However, the enlargement of the temporal scale, as reflected in the duration of long term investments and credits, appears to have levelled off in conjunction with the spatial scale reaching global extension. In the early 19th century the speed of monetary transactions was not much faster than the exchange of goods, since they both relied on physical transportation. With the innovations in communication technology such as the telegraph and telephone in the 19th century, the speed of the monetary system started to surpass that of the material economy. In the last two decades, this development has, with the addition of computer and new communication technology, resulted in an unprecedented monetary trading volume, now surpassing the exchange of goods many times in monetary value.

Uncertainty involved in economic activities increases naturally over time and space. The larger spatiotemporal scales of the economic system are governed by society's monetary institutions that reduce uncertainty of monetary value and ability to transact over time and space. The smaller scales depend ultimately on the same institutional framework, but depend more directly on the monetary market's ability to create transactional instruments in conjunction with the use of available information technologies. I therefore suggest that the institutional framework governing transactions in the economic system at large emerges from the range of spatiotemporal scales of the monetary system.

THE LONG-TERM EVOLUTION OF SOCIETIES AND INTERACTIVE TRANSFORMATION OF ECOSYSTEMS

The introduction of monetary institutions appears to have accelerated the transformation of ecosystems and societies. I will here explore the transformation based on the following three observations of (1) ecosystems as nested and scale correlated organizations, (2) the historical expansion of the range of scales of societies and (3) societies' accelerating impact on ecosystems.

(1) Two central assumptions of ecosystems are:

Nesting. Ecosystems are nested hierarchies of biological processes: For example the processes in a pine needle occur at a faster rate (or shorter temporal scale) and are contained in those occurring at a slower rate (or longer temporal scale) in a tree that in turn is contained in the slower functions of the entire forest [11].

Scale correlation. Temporal scales are correlated to spatial scales: The temporal scale of a forest may extend over several centuries or millennia, whereas the temporal scale of a single tree is years to decades and that of a pine needle may be seconds or days long [11] (Fig. 2).

(2) I make the assertion that the range of scales that societies operate within has increased through human history, as illustrated in Table 2. The range has increased through the addition of smaller spatiotemporal scales at a faster rate than the addition of larger scales. The larger temporal scale has historically levelled off, rarely being extended beyond a few decades, in conjunction with the spatial scale reaching global extension. The smaller scales on the other hand have never ceased to get smaller.

TABLE II

Society's range of spatiotemporal scales

Socio-economic system	**Earliest appearance**	**Range of spatial scales**	**Range of temporal scales**
Hunter-gatherer	100,000 B.C.	$10^{-1} - 10^{2}$m	Day - Months
Early agricultural[1]	15,000 B.C.	$10^{-2} - 10^{3}$m	Day - Years
Advanced agricultural[2]	3000 B.C.	$10^{-3} - 10^{6}$m	Hour[3] - decades
Industrial[4]	1775 A.D.	$10^{-10} - 10^{7}$m	Second[5] - decades
Advanced industrial	1945 A.D.	$10^{-14} - 10^{7}$m	10^{-7} sec. - decades

[1] Subsistent agriculture.
[2] Agricultural civilizations like ancient Rome and medieval Europe.
[3] It was around 1330 AD that the hour became one of twenty-four equal parts of the day [2].
[4] Coal energy and industry applying the scientific knowledge of chemistry, as the smallest unit of control.
[5] Seconds were not applied to time keeping until the late 16th century [2].
[7] Oil and nuclear energy and industry applying the scientific knowledge of physics, as the smallest unit of control.

(3) The human induced impact on ecosystems has overwhelmingly taken place during the last two hundred years and at an accelerating rate, as illustrated in Table 3.

These three observations are brought together in Fig. 2 to illustrate how the spatiotemporal scales of human use of the global ecosystem has changed throughout history. The *thick reference line* represents the spatiotemporal scales of the global ecosystem. This line illustrates the rate at which the earth's ecosystems are being regenerated at varying spatial scales, which is assumed to be stable. The *other lines* represent the rate at which societies are using the ecosystems in the form of natural resources as sources or sinks. Lines above the reference line imply that

TABLE III
Selected forms of human-induced transformation of ecosystemic components in years of quartiles from 10,000 B.C. to mid-1980s [12].

Form of transformation	25 %	50 %	75 %
Deforested area	1700	1850	1915
Loss of terrestrial vertebrate diversity	1790	1880	1910
Water withdrawals	1925	1955	1975
Carbon releases	1815	1920	1960
Sulphur releases	1940	1960	1970
Lead releases	1920	1950	1965
Carbon tetrachloride production	1950	1960	1970

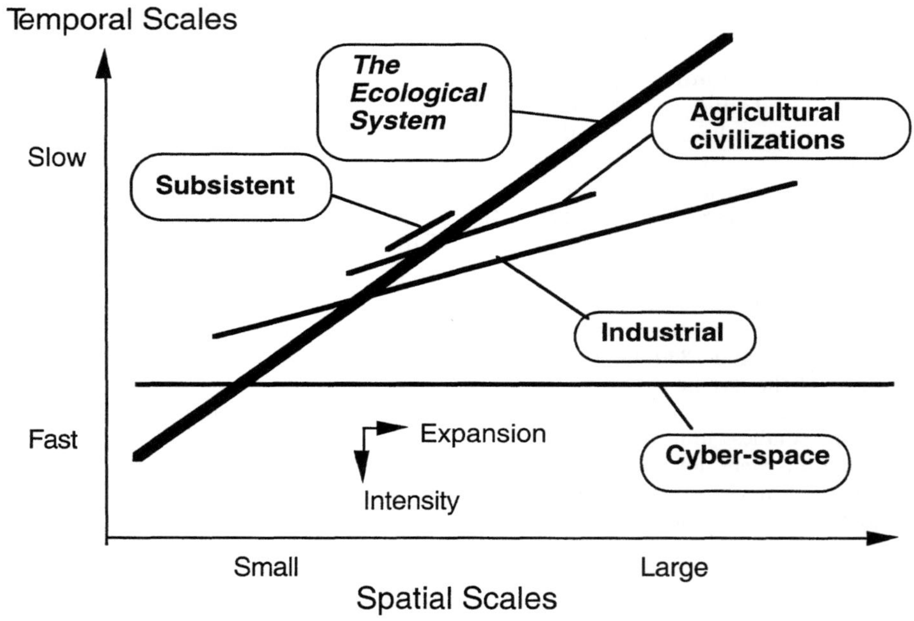

Figure 2. The change of spatiotemporal scales in society throughout history

the ecosystems can replace natural resources being used by societies or absorb waste from societies as a sink; that is, the probability for sustainability increases above the reference line.

The line for *subsistence societies*, whether it is a society of hunter-gatherers or subsistence farmers, follows the reference line closely (Fig. 2). They are closely aligned to the spatial inertia of the ecosystem, which means natural resources are not taken out and used at a faster rate than they can be replaced or absorbed by the global ecosystem. Subsistent societies also interact in a relatively limited range of scales. At this level of development, societies can probably be understood as nested ecological-economic systems.

Agricultural civilizations emerged as the spread of agriculture started to transform ecosystems by increasing the available yield of biomass and the range of scales of society increased (Fig. 2). With the establishment of large scale civilizations, the use of natural resources sometimes exceeded the rate of ecosystem regeneration resulting in ecological degradation and collapse of societies [20], which is indicated by the line crossing the reference line in Fig. 2. These early large scale societies deferred collapse by colonizing new territories as in the case of the Greek civilization. For the first time in history, these territorial expansions introduced the need for monetary systems and large scale trading to act as societal integrators.

Industrial societies took off with the invention of a technology that enabled the use of the energy in coal. The introduction of new monetary institutions appear to have had an essential role in promoting economic transactions in conjunction with new information technologies. This made it possible to increase the speed and capacity of those transactions. Furthermore, industrial societies have been able to increase their spatial scales without increasing their temporal scales, thereby not reflecting the correlation of scales of ecosystems that are being used. This has led to a deviation between the correlation of spatiotemporal scales in which societies operate within and the correlation of scales of ecosystems. Industrial societies have become nested economic systems instead of nested ecological-economic systems.

CONCLUSIONS

In this paper I have suggested that the economic system is a nested organization of spatiotemporal scales and that the range of those scales emerge from monetary institutions. The economic system promoted by the access of information technology has been able to run faster and faster at ever shorter temporal scales. Society appears to be approaching a 'cyberspace-like' economy operating at a global spatial scale and at a close to zero temporal scale (Fig. 2).

This development is in total opposition to the view of the economic system at large being a *de facto* subsystem [7] of the ecosystem. The spatiotemporal scales of the economic system do not reflect the scales of the ecosystem at large. Early societies were largely nested ecological societies. Money as an economic integrator of societies emerged as the scale of societies increased beyond small scale subsistence. Industrial societies have increasingly become nested economic systems. The economic system is the major integrator of all of society's functions. Under its influence, resources or their sinks are used at a faster rate than the ecosystem is able to regenerate them. The gap between the ecological and the economic systems is widening.

A major objective of the monetary system has been to reduce friction of human activities in the ecosystem. It is precisely the unlimited removal of this friction that has led to unsustainable economic development. We have to acknowledge the whole spectrum of spatiotemporal scales of the ecosystem if further depletion of the ecosystem is to be avoided. In conclusion, the monetary institutional framework needs to be designed to promote economic interaction with the ecosystem so that it emulates the spatiotemporal scales of the ecosystem.

H. Thomas R. Mark
Natural Resources Management
Department of Systems Ecology
Stockholm University
e-mail: equinox@algonet.se

REFERENCES

[0] The British 18th century was characterized by strong growth of population and domestic economic activities. There were only about 10-15 county banks in the countryside of Britain in 1750, due to a financial crash (The South Sea Bubble). In 1797 there were 290 county banks, 370 in 1800 and over 600 in 1810 [4].

[1] **Beniger, J.R.**, *The Control Revolution*, Harvard University Press, Cambridge, USA, 1986.

[2] **Boorstin, D.J.**, *The Discoverers: A History of Man's Search to Know his World and Himself*, Vintage books, New York, USA, 1985.

[3] **Braudel, F.**, *Världens Tid (Le Temps du Monde)*, Gidlunds böcker, Stockholm, Sweden, 1986.

[4] **Braudel, F.**, *Marknadens Spel (Les Jeux de L'Echange)*, Gidlunds böcker, Stockholm, Sweden, 1986.

[5] **Coase, R.H.**, *The Institutional Structure of Production*, The Nobel Foundation. Sweden, 1991.

[6] **Coase, R.H.**, "The Nature of The Firm", *Economica Journal* (Number **4**), 1937.

[7] **Daly, H.**, "Alternative strategies for Integrating Economics and Ecology", In: **Jansson, A.M.** (ed.), *Integration of Economy and Ecol-*

ogy. An Outlook for the Eighties, Proceeding From The Wallenberg Symposia, Sweden, 1984.

[8] **Dillard, D.**, *Västeuropas och Förenta Staternas Ekonomiska Historia (Economic Development of the North Atlantic Community)*, Gleerups, Malmö, Sweden, 1967.

[9] **Ekelund R.B. and Hébert, R.F.**, *A History of Economic Theory and Method*, McGraw-Hill Publishing Company, International Edition, 1990.

[10] **Holling, C.S.**, "The Resilience of Terrestrial Ecosystems; Local Surprise and Global Change", In: **Clark, W.C. and Munn, R.E.** (eds.), *Sustainable Development of the Biosphere*, IIASA, 1986.

[11] **Holling, C.S.**, *Cross-Scale Morphological Geometry and Dynamics of Ecosystems.* Ecological Monographs, 1992.

[12] **Kates, R.W., Turner, B.L. and Clark, W.C.**, "The Great Transformation", in: **Turner, B.L.** (ed.), *The Earth as Transformed by Human Action*, Cambridge University Press, USA, 1990.

[13] **Kennedy, P.**, *The Rise and Fall of the Great Powers*, Vintage Books, New York, USA, 1989.

[14] **North D.C.**, *Economic Performances Through Time*, The Nobel Foundation. Sweden, 1993.

[15] **North, D.C.**, *Institutions, Institutional Change and Economic Performance*, Cambridge University Press, USA, 1990.

[16] **O'Neill R.V. et al.**, *A Hierarchical Concept of Ecosystems*, Princeton University Press, USA, 1986.

[17] **Rosenberg, N.**, *Perspective on Technology*, Cambridge University Press, USA, 1976.

[18] **Rosenberg, N.**, *Inside the Black Box. Technology and Economics*, Cambridge University Press, USA, 1982.

[19] **Schumpeter, J.A.**, *The Theory of Economic Development*, Cambridge: Harvard University Press, 1934.

[20] **Tainter, A.T.**, *The Collapse of Complex Society*, Cambridge University Press, USA, 1988.

[21] **Vilar, P.**, *A History of Gold and Money (1450-1920)*, Verso, London, England, 1984.

[22] **Weissmahr, J.**, *An Evolutionary Theory of Technical Change and Economic Growth and Development*, The European Study Group for Evolutionary Economics (ESGEE), Zürich, Switzerland, 1991.

RICCARDO PETRELLA

THE POST-INDUSTRIAL SOCIETY: WHAT FUTURE ARE WE PREPARING?

INTRODUCTION

In this paper we will submit the following thesis to the reader's reflection. The crisis of the paradigm of the industrial society that has manifested itself since the mid-seventies, has favoured since thirty years the promotion and dissemination throughout the world of two new global paradigms: the paradigm of *information society* and the paradigm of *sustainable development.* Both new paradigms, having quite different roots and aims, have developed quite independently the one from the other, i.e., without clear-cut opposition nor convergence occurring.

Since the mid-eighties, the information society paradigm seems to have outgrown the sustainable development paradigm, as well on the rhetorical, on the economical as on the political level. Ever since, the information society paradigm has definitely become the most powerful and best accepted regulator of contemporary societies, the future of which it is shaping. My key-argument is that this success is due to the fact that the information society paradigm is *not* radically new. There is a great continuity with the *industrial society* paradigm. As a matter of fact, it is a type of neo-industrial modernisation paradigm that, in spite of its different techno-economic foundations and its completely new spatial framework (the global has replaced the national), has not changed at all its nature, that remains capitalistic. In a certain way the information society paradigm is a fulfilment or an amplification of capitalism. The sustainable development paradigm from its side is grounded, originally at least, in concepts, principles and aims quite in contrast with those proper to the industrial society paradigm.

The global information society will not fulfil its promises anyway. It will not pave the way for a new society reconciling social justice, economic efficiency, political democracy, cultural diversity and sustainable development. Such a reconciliation will only be feasible and solid in the presence of a social contract at the global level. There will even be no future for industry itself, in the absence of such type of contract.

© 1999 *Kluwer Academic Publishers. Printed in The Netherlands.*

TRIUMPH AND DECLINE OF THE INDUSTRIAL SOCIETY AS A DOMINANT PARADIGM

During the whole period starting with the industrial revolution from the 18th century onwards, over its explosive expansion in the last century and including its triumphal growth after the second World War, *industrial modernisation* has been the primary mover and the cultural and political metaphor of our societies. Since it has become universally applicable, this process has been disseminated throughout the world in the colonial era and even after decolonisation.

Until recently, industrial performance indicators have been generally accepted as adequate tools to define and describe innovative enterprises, regions, societies and countries all the like. *Industrial growth* has become the paramount symbol of the *industrial society.* A general conviction has grown, that scientific and technological progress, applied to new industrial processes and products, would automatically lead to social well-being and economic 'progress'. Contrariwise, this economic 'progress' was considered to be a *conditio sine qua non* for social well-being and social progress. No social 'progress' (and, consequently, no democracy), without economic growth! No economic growth without technological innovations in industry! In this context, those innovations had to be, in the first place:

- aimed at larger quantities (more steel, cars, radios,...),
- productivity-centred (expressed in monetary terms),
- dictated by considerations of short-term efficiency of the enterprise and through competition on the market.

The fundamental objective of research and development (R&D) was to increase the efficiency (energy, natural resources, base-products, labour -force and equipment...); the categorical imperative was to lower the costs in monetary terms (hence competition of prices). The guarantee of 'full employment', paralleled with an elaborated social system (social security, pensions, equal opportunities, anti-poverty measures), together with access to mass consumption products and services were society's offers to those who had a job.

In other words, it is thanks to this type of 'social contract' (the welfare-state) that the industrial modernisation, originated and implemented in the industrialised nations, was made possible and came to full growth. In the seventies and early eighties an important change has occurred in parallel with the use and dissemination of a new generation of technologies (micro-electronics, computers, telematics,...): a displacement from *serial production* ('taylorism' and 'fordism') to flexible *specialised production* and *personalised production* ('toyotism') (see Table 1).

TABLE I

Main characteristics of serial and specialised flexible production

"taylorism"	"toyotism"
- standardised serial production	- sophisticated quality products
- limited product variety	- customer tailored product variants
- single issue technology	- small production units
- poorly qualified labour	- programmable technology
- thoroughly divided labour	- limited division of labour
- major production units	- skilled labour
- well structured organisation	- bureaucratically developed production factors
- limited R&D activities	- high salaries
- price competition	- intelligent organisation
	- quality competition

Source: Commission of the European Communities, FAST II: Program results and Recommendations, Brussels, 1988.

This displacement has been fostered through a process of a *high-tech industrial modernisation* and through the emerging *neo-industrial society*, characterised by:

– a quality-oriented innovation,
– a substantial increase of the non-material costs (R&D, training, maintenance, publicity,...), as compared to the overall production costs,
– the establishment of regional free-exchange areas and economic unions at a higher speed,
– an increase of competition amongst corporations, regions and countries at the international and global levels.

Nevertheless, in the same period the main driving force (growth) and the fundamental principles (short-term efficiency, continuously increasing productivity, technology-based innovation) of this same paradigm of *industrial society* have started to come under criticism and severe attack. At the end of the seventies and in the early eighties it was generally acknowledged that the paradigm of *industrial society* had no more future, for reason of the existence of two major structural limits.

The first limit lies in the fact that it was no longer capable to provide stable and solid foundations for the national welfare-state. The continuous increase of productivity and the growing competition, due to the introduction of new technologies on the ever more global marketplace, have caused the employment crisis, in spite of great efforts in favour of job-creation and of humanisation of labour in Germany, Japan and in the Scandinavian countries.

In addition, the more *industrial society* tried to combat the high unemployment rate by means of an intensification and a higher gear of industrial modernisation of the high-tech type, the more this unemployment became progressively more structural. This has resulted in an explosion of old and new forms of poverty and social exclusion, which in turn has occasioned the rise of public debt and the crisis of the welfare-state. Ever since, the important restructuring of the welfare-state has destabilised the 'social contract', on which the paradigm of *industrial society* had been built and acquired its social legitimation since World War II. It had become evident that industrial society 'had progressively lost its capability to support the social contract'.

The second limit is related to the internal logic of the industrial modernisation paradigm, with its intense consumption of renewable and non-renewable resources. This excessive use has caused a massive environmental destruction and a deterioration of the ecological dynamics of our planet.

Billions of tons of solid waste, essentially the product of the advanced nations of the northern hemisphere (especially the USA.); the evacuation and destruction of mountains of extremely toxic wastes; a critical degree of soil erosion; deforestation and desertification; the death of rivers and lakes; progressive degradation of coastal areas and the lives of their inhabitants; the pollution of the seas to great depth; climatic warming (it is impossible to review the list of ecological catastrophes already occurring, and the environmental threats and forecasts about the environment of future generations, without shivering). All this has transformed industrial modernisation into real *ecological barbarism*, and extremely so in those industrialised countries where the paradigm of *industrial society* has in no way been mitigated or tempered by any ecological policy (as is the case in the former soviet-Union and the 'socialist' countries of Eastern Europe).

NEW PARADIGMS THAT COULD POSSIBLY REPLACE THE PARADIGM OF THE INDUSTRIAL SOCIETY

Since the publication of *The coming of Post-Industrial Society* by the American Daniel Bell in 1973, that coined and popularised the idea of a *post-industrial society*, many have proposed sharp definitions for this new concept (see Table 2).

Two of these proposals have acquired fame, esteem and adhesion all over the world: I have in mind the paradigms of *the information society* and of *sustainable development.* In contrast with other paradigms, that have received a marginal support (i.e., geographically, temporarily or

TABLE II
Would-be substitutes for the industrial society paradigm

service society	society of know-how
post-capitalist society	innovation society
conservation society	information society
sustainable development	post-materialist society
knowledge society	communautarian society
post-modern society	informal society

Source: R. Petrella 1996

socially limited), those two have become virtually universal in terms of credibility and support. Apparently they are the cornerstone of a new radical turn we are now involved in.

THE PARADIGM OF SUSTAINABLE DEVELOPMENT

Known in the sixties and seventies, although with other names, the concept of *sustainable development* has been successfully launched by the Brundtland Commission in 1987 (thus proving that commissions can be capable of inventing useful and influent concepts!). This concept has had a tremendous impact on ideas, on ways of defining problems, on values, on strategies and behaviour of people all over the world.

Generally speaking, the following ideas have gradually made their way:

- economic development has to respect the environment, and should not create deleterious conditions for the generations to come,
- R&D, industry, finance, consumers and political authorities in the northern hemisphere should give priority to innovation in order to reduce the environmental destruction,
- science and technology should be directed to conceive, develop, produce and implement new and environment-friendly production processes and services,
- in the southern hemisphere, the most urgent is the struggle against poverty and environmental destruction.
- to implement principles of this type, shallow modifications or minor corrections to the rationale of industrial modernisation are far from sufficient. The paradigm of sustainable development calls for an in-depth transformation of the principles and mechanisms of industrial society, in the 'advanced' countries and in the poor countries alike, with due respect to:
- deep change of social and economic values, with concomitant mutation of life-styles and consumer behaviour,

– deep modification of the fragmented philosophy and education of engineers,
– selection of a new set of socio-economic indicators and accountant systems,
– transformation of the processes of decision-making, legislation and power-legitimisation of governments and empowerment of transnational authorities and agencies, based on global standards (see the new *World Forestry Stewardship Council*),
– a new vision on corporations and a new approach to management, in the world of finance to begin with (not exclusively 'green capitalism'!),
– a restructuring of the education system for the long term, at all levels.

The paradigm of *sustainable development* has paved the way for a great number of important innovations. It would take us far too long to enumerate all new legislative initiatives, new laws and decrees that have been voted and introduced in almost all countries of the world, in virtually all economic sectors existing, and related to this new paradigm. The same could be said about the endless list of action plans and charters for the environment that have been approved, and are now under way of being implemented at the local, regional, national, European, international and global scale. Doubtless, Agenda 21 is the most well-known at the global scale. It has been endorsed by 130 heads of state in Rio de Janeiro, 1992. Nevertheless, this plethora of laws, proposals, action plans and charters notwithstanding, the deterioration of the state of the environment is not halted, neither at the global nor at any other level, as it seems (apart from a substantial number of positive exceptions). Why?

The reasons for the deterioration are manifold; they vary from country to country and depend on the economic sector concerned. My personal point of view is that one of the main reasons is to be found in the current impact, popularity and rising support to the paradigm of the *information society*, together with the globalisation of the economy and society under the iron rules of liberalisation, dereglementation, privatisation and competition.

THE PARADIGM OF THE INFORMATION SOCIETY

In 1972, a first detailed plan for the transformation of an 'industrial society into an information-based society' has been published. It is called *Plan for Information Society. A National Goal Towards 2000*, and has been commissioned by the Japanese government. The main objective of the plan was a redistribution of industrial activities throughout the

Japanese territory, thanks to the establishment of a new information and communication network. Another objective was to install an environmental survey and prevention system. According to its authors, the information society should foster:

– an important rise of the quality of life in Japan,
– a substantial increase of creativity of the people,
– a great impulse for participative democracy,
– an intensely convivial society.

Only a few years later, President Giscard d'Estaing took the initiative for the French plan 'L'informatisation de la société', that declared information and communication technologies on the verge of radically transforming the industrial society. The *information society* would favour a new growth and a new power game. National sovereignty was under challenge. It is an impossible task to quote the endless list of papers, books, conferences, discussions and reports relative to the *information society* in the 60's and 70's in the USA. Suffice it to mention the best-seller by Alvin Tofler *The Third Wave*, published in 1981, that contributed to disseminate the idea to millions of enthusiastic readers all over. In the USA, after having read this book, businessmen, public administrators, politicians, educators and professionals of the media all alike were deeply convinced that the ongoing transformation to an information society, represented no lesser mutation in history of humankind, than the agricultural and industrial revolutions had been in the past.

However it may be, during the 70's the industrialised countries have been seen to adopt a general consensus on three beliefs:

– A new era had started, the era of the information society;
– The information society is characterised by:
  - higher living standards
  - new growth possibilities
  - new forms of labour
  - better perspectives for man's creativity
  - universal access to knowledge and education
  - explosive growth of interactivity and conviviality
  - more democracy
  - a more stable peace
  - a better and more effective prevention, survey and rehabilitation of the environment;
– The *transition towards the information society* became the prime objective. Such was the situation about 20 to 25 years ago.

From the point of view of the seventies, clearly the *information society* did not keep its promise. Faced with its prophecies, declarations and projects the true accomplishments remained rather modest and extremely disappointing. The most remarkable achievements in the history of the information society of these last twenty years concern the breakthrough of informatics (data collecting, processing and transmission) in terms of speed, quantity and quality, reliability, miniaturisation and cost reduction.

The great changes happened in other domains, those of production. Information and communication technologies allowed the effective transition from 'fordist' serial production to 'toyotism' and its fine-tuned quality production (see Table 1). An interesting question concerns the development of the information society in the domain of health, leisure, education, formation and media, where new 'markets' have opened up.

Beyond this, my feelings are that despite the fascination, enthusiasm and the enormous hope aroused by *the information society*, the new paradigm did not succeed in reducing the inequality between people, groups, cities, regions and countries on the contrary; the information society has magnified disparities and favoured new processes of recentralisation of power and concentration of riches. It did neither contribute to transform cities into peaceful places of coexistence and co-development. Since a few years, the information and communication technologies are associated to the creation and unfolding of so called *private cities*, a kind of 'exclusive cities' inside of the cities.

In contrast, the conception and development of the applications of the information society have been dominated by rules of the market and by cost effectiveness of concerns and a short term gain of financial power.

Finally, the most important contributions of the information society have been:

a) developments of military equipment offshoot;

b) an acceleration and amplification of the globalisation process of the economy and, therefore, a liberalisation, deregulation and privatisation process of the market, closely linked to the former.

But, how to explain this new wave, more intense even than before, of fascination and enthusiasm that brings about today the paradigm of the information society, whereas a quarter of a century of history of this information society clearly has shown the enormous distance separating projects, prophecies and promises from true realisations and real changes in economy, politics, quality of life and society?

The essential reason for this madness lays without doubt in the second generation, today 10 years of age, of the 'revolution of information technology', the one associated with the 'digital revolution'. Digitalisation represents effectively a major technological discovery. It reshapes profoundly the foundations and configurations of the worlds of the media and information and communication systems. The possibilities offered by digitalisation to integrate on the same carrier voice, images, data and text has given to the computer and telecommunication industry, to television and the press, to leisure industries and show business a boost of energy. The possibility of opening up new large markets and to revitalise the applications of the information society to manufacturing, health, banking and insurance, public services, education, family leisure or transport gives to financiers, industrials, business people and politicians the feeling that the information society really knocks at the door and that the most urgent and useful task is to be receptive to this second era of digitised information.

Quite a few authors, notably Tofler (*The New Politics in the age of Information Society*) and Negroponte (*Be digital*), have given to the 'renewed enthusiasm' intellectual and techno-scientific foundations. It has become increasingly difficult to follow the whole *new* series of national plans (see Table 3) elaborated and approved in many countries of the North and South. These are inspired and guided by the same ideas and principles, more precisely that the information society:

- eradicates all geographical and political frontiers. The information society can only be founded on the principle of a complete liberalisation of the global market;
- remodels the spirit of competitivity, replacing a competitivity based on price reduction by one of competitivity of treatment, of quality and the speed of information;
- will remould all public and private spheres;
- will create a new organisation of labour that will give priority to intellectual work and professions concerned with information;
- will destabilise the traditional hierarchical relations and structures, giving access to new, more advanced forms of participative democracy;
- will cause a new era of growth and economical wealth.

The *Europe and Global Information Society* report, drafted for the use of the European union by the 'Bangemann' group, proposes the stance in favour of the information society in its most loyal and enthusiast expression, the ideas and principles conform they way we enumerate them below.

The appeal of technology is evident: people, attracted by everything of technocratic nature, are delighted to be able to exploit new powers of

TABLE III

Planning for an information society

Country	Year	Title
Denmark	1994	Commission for the information society, *Info-Society 2000*, Copenhagen, Ministry of Research
France	1994	Théry Report, *Les Autoroutes de l'information*, Paris, La Documentation Française
Netherlands	1994	Action program for the electronic highways, *Information Super-highway—From Metaphors to Action*, The Hague
United Kingdom	1994	Commerce and Industry Committee, *Optical Fibre Networks*, HMSO, London
Sweden	1994	Governmental commission for information technology, *IT-wings to Human Ability*, Stockholm.
European Union	1994	Commission of the European Union, *Growth, Competitiveness, Employment*, White book, Brussels
	1994	"Bangemann Report", *Europe and the global Information Society*
	1994	Recommendations to the European Council, Brussels, *Europe towards the Information Society. An Action Plan.* Communication of the commission to the European Council and Parliament, Brussels
Australia	1994	*Networking Australia's Future.* Final Report of the Broadband Services Expert Group, Sidney
Canada	1994	Proceedings of the Congress of the Consulting Council on information highways, *Canada's Information Highway. Building Canada's Information and Communication. Providing New Dimensions for Learning, Creativity and Entrepreneurship*, Ottawa
United States	1993	NTIA, *The National Information Infrastructure, Agenda for Action*, Washington DC,
	1995	*Global Information Infrastructure, Agenda for Cooperation*, Washington DC
Singapore	1992	National Council of Informatics, *A vision of an Intelligent Island: the IT 2000 Report*
Thailand	1995	*IT 2000. Towards Social Equity and Prosperity.* Thailand's IT policy for entering the 21st century, Bangkok

transformation of the economy due to:

- reorganisation;
- concomitant organisation;
- the global system of reorientation of production;
- supplying on world scale.

The appeal exerted by the market is, by itself already, considerable. If I were a business man and, more particular, a chief manager of one of the major multinationals in the industry of telecommunication, computers or media, I would certainly be excited by the perspective of making huge profits and strongly inclined to enter into this new competition and new global 'war' in the hope of expanding my global market share, increasing my capital, enriching myself and controlling a part of the greatest information machine of the world.

This new fascination and enthusiasm are fruitful in such a measure that the paradigm of the information society slowly creates a vacuum around itself. It has become the prime priority of all organisations, public and private; it represents to our society *that* future we are anticipating and aspiring; it hides behind all terms of contemporary political rhetoric, it lurks in the 'let's go for it' or 'we better catch that train'.

The paradigm of the information society submits or absorbs all other dimensions, all other questions, all other perspectives. As shown by the famous *White Book* of the European Commission on 'Growth, competitivity, employment', the development of the information society is considered the condition *sine qua non*, the key-process to allow the restoration of the economical growth in Europe and to admit her to enter the 21st century. The role of the paradigm of *sustainable development*, to which the *White Book* largely pays attention, seems rather marginal in comparison to the dominating current, i.e., in comparison to the information society. The fascination and enthusiasm, by themselves, do not explain how 'information society' has been able to bypass 'sustainable development' in the intellectual and socio-cultural domain, as well as on the level of political lines of action.

The second major reason for the grip due to this paradigm is explained by the ideological function and the economical role the information society fulfils.

As we have stated above, starting in the mid 80's, we have continuously increased the call for information and communication technologies in order to justify and accelerate the process of liberalisation, deregulation, privatisation, globalisation and encouragement towards competitivity.

If, in fact, the acceleration of the process of globalisation of financial mobility following Nixon's decision in 1974 to liberalise the movements of capital from and towards the USA, and the globalisation of financial markets that resulted have been that fast, extended and intense, it is thanks to the new information and communication technologies. The global flux of *electronic* currency is steadily redefining and restructuring the *global system* and the role of finance in the *real* global economy.

Thanks to the information society, the globalisation of financial mobility and financial markets has without doubt intensified and augmented that of manufacturing of goods and the creation of services, it has accelerated the globalisation of companies and their strategies, that of markets, consumption models and society in general.

In such a context, every new break-through in information and communication technologies justifies and 'legitimises' the forces in favour of the three principles:

- *privatisation*: the common argumentation is that the information society renders the information increasingly accessible to everyone and favours the dissemination of new ideas, the development of a new entrepreneurship and new markets. All the activities therefore have to be entrusted to private property, to private politics and leadership;
- *liberalisation*: the argumentation put forward on different occasions by the *Financial Times, The Economist, Newsweek*, and all over the world in different papers of 'liberal' signature, is the following: the more wealth will depend on knowledge and information, the more effectively the prospected advantages will be harvested, under the condition however of liberalising the international markets and abolishing all barriers, protections and regulations;
- *deregulation*: the proposed argumentation states; it is not up to the government to define the principles that control economical activities. This role has to be entrusted to the forces of the market, as technology of information and telecommunication can deliver, more adequately than state 'bureaucracy', the goods and services capable of transforming banks, insurance, health care, education, environmental management, etc.

The foregoing clearly shows ideological, cultural, political and economical premises and foundations of the paradigm of the information society coincides largely with those prevailing in the actual economical system, i.e., the liberal capitalist economy. It seems reasonable to think that remarkable convergence exists between the global information society and global market capitalism. More and more they appear as two faces or as two modes of development of one same reality.

Without being either acceptable nor justified, the grip, the enthusiasm and the credibility evoked by the paradigm of the information society are not surprising.

The Bangemann report, earlier mentioned, identifies clearly and explicitly this convergence. This important political document recommends the political leaders of the member countries of the European Union to put their fate in the mechanisms of the market and the private domain, and to leave to these the care of opening the way that will lead the different European societies to one single information society.

FUTURE PERSPECTIVES: BEYOND THE CONTEMPORARY INFORMATION SOCIETY

The contemporary evolution of the information society does not give certitude nor sufficient guaranty to allow to think that the future will be better concerning the following:

- more social justice, less inequality;
- a more performant organisation of the economy;
- a larger civilian participation and a more effective democracy;
- a society more respectful towards cultural diversity;
- a globally more sustainable ecological development.

I do not pretend that the information society will not, in principle, be able to contribute to the construction of a promising future. What I say, is that the concepts, the principles and the rules from which *today* the information society draws its inspiration, will *not* contribute to the creation of a better future.

By using the new possibilities of the information and communication technologies, one could more easily construct such a future, by according priority to the elaboration, the definition and by carrying into effect a *global social contract* that could:

a) redefine and establish the common global public cause (the global interest) based on the production and valorisation of the common global resources (water for example) and global services (like education, health, basic telecommunication), and

b) installation, on the global level, of new institutions and new organisms that would create a co-operative system able to re-establish a political control over infrastructures and financial and economic activities safeguarding the interests of the whole world population.

The future does not reside in a globalised information society, liberalised, deregulated, privatised, subjected to competitivity installed as a supreme regulator. It does consist of a sustainable development, social and global, relying on a 'global social contract' that will allow all social groups, all countries and all regions of the earth, to profit more fully of the infrastructures, networks, products and services originating from the new information and telecommunication technologies.

Riccardo Petrella
President of the 'Group of Lisbon'
Commission of the European Community
Brussels, Belgium

JANINA ROSICKA

SCARCITY VERSUS FERTILITY TWO WAYS OF THINKING ABOUT ECONOMY

INTRODUCTION

An idea has creative power. We see what we want to see. However, what we want to see is dictated by our ideas. Postmodernism attempts to widen our sight. Hermeneuts, Foucault, and Derrida want to cure people of naive realism. For them history is not dead. It is rather a kind of equivalent to the scientific laboratory. As such, history enables us to test different solutions invented in the past. History helps us to grasp things long forgotten. This function creates an opportunity to follow the way from human emotions to their rationalisations, and in consequence an important opportunity to learn something about human arrogance. In 1953 an economist W.S. Vickrey wrote: '*Economic theory, in its purest and most abstract form, can be treated as a system of logic, having no more immediate ethical content than a proposition in Euclidean geometry*' [27]. The majority of economists are still thinking about economy as a logical play in which people calculate their activities with the accuracy of a robot. My proposal is to seek an antidotum to this attitude in the history of economic thought. This social science shows that in the past people used to think about economy in quite a different way.

When people of the Enlightenment decided to judge humankind by the Supreme Court of Reason, they did not know that in return a culture of postmodernism would be the judge of the Enlightenment ideas. Our contemporary philosophy accuses rationalisation promoted by the Age of Reason, of creating the chaos of modern world. As Karl Jaspers said, the main power of history is the rationalized madness of frightened humankind [7]. This judgement means that humans as irrational beings use inappropriate tools, and history is a testimony of their activities. I would like to make Jaspers' statement the starting point of my reflections. I will look for an explanation in the cultural context of economic ideas of the following problem: *what kind of human madness was rationalized by economic ideas and what was the impact of this economic rationalization on the thought about Plenty and Scarcity?*

© 1999 *Kluwer Academic Publishers. Printed in The Netherlands.*

PLENTY AND SCARCITY

The ideas of Plenty and Scarcity were and are two very important economic ideas. When we have many goods, the problem is how to deal with and how not to waste those goods. However, we can fear plenty. When we think that we have a lack of goods, the problem is how to overcome that scarcity and we are permanently afraid of scarcity. In the first case we consider economy as a potential but not very common danger; in the latter we consider economy to be an eternal cause of danger and frustration. In our history, plenty was connected with ancient and medieval 'economics' of family, and with socialism and Marxism, whereas scarcity is rather a category of the state analyzed by the eighteenth century Political Economy and then by economics. Modern economics link creativity, justice, competition and productivity in a single set of values, conditioned by scarcity. Those links have become a kind of moral blackmail and they serve to justify every sort of inequality. So the moral sense of scarcity is inequality and, contrary to this, the moral sense of plenty is equality. Following Adam Smith, modern economists claim that plenty as a goal of economy resulted in poverty, and scarcity as the main goal of economy has resulted in welfare. In this way, scarcity became the value which marked out the area of an interest in economy and economics by describing the main feature of economic fact. As the reverse of scarcity, plenty became a feature of non-economic fact and it passed out of the sight of economists.

FERTILITY AND THE AGRARIAN MYTH

When we open a handbook of economics we see scarcity as the first word. When we try to read an economic view of primitive magic and religions the first word in that context will be fertility. Present-day economists try to cope with the scarcity of productive agents. The primitive 'economists' had to ask the gods for fertility. Thought about economy was organized by the agrarian myth. Fertility was the gods' power. Inseminated by them, the Earth (the land) supplied the means necessary for life. Primitive communities knew what they needed. Their material needs were stable and they primarily embraced food. The result of this stability of needs was economic equality. In the course of time, the town and state broke this equality. However, they could not break the ancient mythical imagination. Christianity had to adapt the old myth and medieval 'economists' dealt with the problem of subordinating the town to the country pattern. They invented the institution of the *just price*, which limited the expansion of the town. The destruction of the agrarian myth was initiated by seventeenth and eighteenth century trends:

the Jesuits' economic doctrine and mercantilism. Then, in the fifties of the eighteenth century French economists (named *Physiocrats*), wanted to deliver the old myth through its rationalisation. And finally it was David Hume who introduced the word 'scarcity'. The father of the political economy, Adam Smith , supported Hume's view and he recognized the old myth as barbarity. In mythical fertility (he called it opulence or plenty) he noticed an opposite thing—scarcity. Smith was a moderate optimist and he thought that a division of labour and productivity would reduce economic scarcity. His followers, Malthus and Ricardo , did not share his optimism. They combined an admiration of industry with a pessimism of agricultural opportunities. In the second half of the nineteenth century the political economy gave way to the Austrian School and Marshallian economics. Scarcity became the basic economic category, the main feature of economic fact. In contrast to these trends, Socialism and Marxism continued the physiocratic approach—the rationalisation of the agrarian myth. Saint-Simon, Fourier, Proudhon, and Marx wanted to animate the old pair 'fertility-plenty' and instead the power of the Earth they proposed to use the powers of human labour.

Now neither theory nor policy can cope with our contemporary 'plenty', known as overproduction or unemployment. The replacement of the magic pair of fertility-plenty by scarcity resulted in a deep change in our thought about economy. Economists use to think about scarcity as an advantage. Creating a new need is better than satisfying an old need, for example by reduction in prices. In the nineteenth century, human needs were a main source of economic energy; at the end of the twentieth century they must be supported by advertising. The modern economy replaced the 'aggressive generosity' of the primitive community with 'aggressive advertising'; a constant Vanity Fair took the place of the primitive buzzing feast. I would like to show how this has happened.

The attitude of primitive man toward economy was determined, like his other activities, by myth, mainly by the agrarian Earth-Mother myth. Man and nature fused together into a living organism [1a]. Seeking harmony and rejecting change; these two trends strengthened the main value, perfection, understood as an absence of change. For economy the lack of changes meant a demand of autarky. Economic self-sufficiency of the autarkic economy could be disturbed by a serious shortage, or by too great a surplus. The community was able to cope with a natural calamity by changing their places of stay, or by getting food from other communities, not necessarily by voluntary exchange. From the point of view of the community, more problems were caused by surplus, because surplus created different challenges. In this case there were inventions like feasts, and burying the surplus in the ground (for example Scandinavian treasures [3a]. Exchange of gifts and feasting created a 'market', within which there was strong competition [3b] [4] [5]. Aron Guriewicz named it the *competition in generosity* [3c].

Both magic and religion initiated the creation of economic terminology. The community's members tested the efficiency of their invention by using magical means. In a magical language they expressed their wishes. Magic was a special sphere within which they formulated their problems, and it was a special sphere for elaborating technology. The centre of economic activity was a cult of fertility. The subjectivity of man, lack of alienation, 'warm' social ties, an emotional connection with the object of working the land; all these features had combined to form an infrastructure for the economy [1b].

In spite of the growing number of merchants and towns, the economy was still perceived through the agrarian myth. Magic economic practices were incorporated into Christianity (the decline of magic is dated in the seventeenth and the eighteenth century [26]). More and more frequent economic crises of famine were rationalized. People regarded them as signs of God's anger and the necessity of penance. Apart from the still active tradition of feasts, the surplus was now used for building churches and to support beggars. The Lord kept the economic balance by using this punishment and by receiving tributes.

However, by its doctrine of individual salvation Christianity introduced, strangely for the pagan world, individualism. Economic growth separated the rich from the poor. At the beginning this division was identified with the division into sinners and good Christians. Soon, however, Clement of Alexandria and Christian writers set their wits to a question: how can a rich man find salvation?

In summary, there were three phases in Christian thought about economy. In the first phase Christianity brought about a revival of the primitive community's values. Next, there was an effort to subordinate town economy to country economy. During the third phase, the town and then the state gained some kind of emancipation [23]. The Church was unsuccessful in its efforts to stop the flow of time. The era of scarcity was near and the problem of fertility and generosity became less and less important. In the Renaissance Machiavelli said the wise does not care about his reputation as a miser. Generosity was necessary in the period of the prince's efforts for power only. Then, when he became the prince he had to be more careful with money [11]. In the medieval Thomistic language money was *sterile*, for pre-modern 'economists' (called mercantilists) money was *fertile*. Mercantilism was the first attempt to revise an agrarian myth. Mercantilists wanted to replace the power of the Earth-land with the power of money, and by the power of productive labour.

Instead of the magic Thomistic universality of Mother-Earth myth they proposed a new aim for economy; wealth of state. The mercantilistic discussion was very long and it lasted till the Enlightenment.

In the sixties of the eighteenth century, François Quesnay , the founder of Physiocracy , returned to the agrarian myth. Physiocracy was the first complete economic doctrine describing the end of a period full of worry about food and the beginning of the industrial period. Quesnay was a scholar who attempted to combine both an old human myth of agriculture and science. He became a mediator between popular imagination (strongly influenced by myth), and new modern scientific concepts. This unity of the old myth and science brought forth a very expressive and suggestive doctrine which directly or indirectly influenced the imaginations of the Enlightenment writers. The power of Physiocracy became a myth. Science was a methodological tool, a test rather than a main source. Science as a testing tool was implementing the myth. This interpretation of the myth of agriculture and especially of its attributes, Regeneration and Fertility , gave rise to modern economic terminology. Reflection on the use of those powers created a modern theory of economics. Quesnay wanted to understand the errors of a mechanism which—as a design of God—should be perfect. So he asked these great questions, which still interest economists. His conviction that agriculture is a perfect mechanism became a source of theoretical economic thought. Looking for reasons for failure produced a different analysis of economy—quest for the state of equilibrium. Seeing agriculture as a reservoir of energy, he attempted to analyse the social structure and the rules of its organization.

Quesnay paid attention to the physical and biological attributes of land. He saw them in the same way as they were defined by myth. He accepted the main requirement of the myth; if we satisfy the demands of land, if we—using the mythical language—do not harm land, the powers of Regeneration and Fertility will reward us in abundance. This requirement became the basis of his considerations. In each of his paragraphs Quesnay tried to answer a question: how to satisfy the physical demands of land in order to achieve welfare.

Quesnay did not dream about coming back to the state of nature. He wanted to extend features which made agriculture the exceptional sphere of human activity to industry and trade. His strong belief in the mythical forces of Regeneration and Fertility did not allow for attribution of similar characteristics to other spheres of economy. Quesnay put forward a solution comparable to Leibniz's pre-established harmony, where the Earth performed the role of a tuning-fork measuring 'proportion' belonging to each class both in production and in consumption. One must listen to the language of the Earth, and in the dialogue with her find the rules of welfare. He recognised the state as a great enterprise using resources of the land and bringing profit. The state existed in order to realise the value of its part of the Earth and through foreign trade joined the general cooperation. Everybody was co-depended. Cooperation is one great interdependence and the Earth is the only reference point. Forces of Fertility and Regeneration wait for us, for our reasonable cooperation. The Earth is the base of cooperation for people. Knowledge about it should be widely popularised [22].

Physiocracy ended the epoch of the agrarian myth. Physiocratic rationalization did not become helpful. In fact, magic fertility, Aristotelian 'average' property, Christian harvest—these all meant the approval of mediocrity. It was, for members of a community, for the ancient and for medieval people, a 'normal' (using Foucault's terminology), average, and proper measure. The unit of this measure was food. There were some 'natural' limits of this kind of wealth—its material volume and its propensity for spoilage. The problems were created by things which did not possess these 'natural' limits—the landed property and money. The common opinion was that nature gave necessary things very easily and fertility was reached quite easily, whereas rare things are not necessary for life. In this way nature was understood by Stoics and in this way nature was designed by Christian Providence. This generalization concerning the generosity of Nature and her self restraint can be called the general pre-modern economic principle of plenty and scarcity. An understanding of plenty as fertility describing society where each man was not hungry, and the relationship between nature and man, within which each fruit of Nature collected by man was not wasted. So the perfect economic specimen had to be a self-sufficiency economy governed by the virtue of moderation, and a ban on emulation.

THE ENLIGHTENMENT INVENTION: SCARCITY

The Enlightenment inherited the question of scarcity and plenty. John Locke repeated the two rules concerning the ideal society. He put stress on the ban of wastefulness as the only condition limiting economic activity, and he said that man disturbed the public good if the things which belonged to him were wasted. In his interpretation, *lasting things* made it possible to achieve a fortune. Money was for Locke *tacit and voluntary invention* which created the opportunity to possess *rightfully and without injure* more land than a man could use. The novelty of Locke's definition was his attitude towards luxury. He did not want to follow his predecessors in limiting luxury and calling for modesty in his requirements. For him, it was a false problem. We should not prevent luxury, we have to prevent waste of the public good [10].

The most iconoclastic against the 'natural' fertility was David Hume. He wrote in his *Treatise on Human Nature* the following: *Of all animals with which this globe is peopled there is none towards whom nature seems, at first sight, to have exercised more cruelty then towards man, in the numberless wants and necessities with which she has loaded him, and in the slender means which she affords to the relieving these necessities* [6a]. Nature released human selfishness: *the selfishness of men is animated by the few possession we have in proportion to our wants* [6b]. Because of her scarcity, nature is selfish herself and forces man to work. Nature makes up for human imperfection (caused by *the numberless wants*) by attracting his attention to morality. In this way justice becomes simply a function of scarcity. As Quesnay began from charity of land understood as surplus, so Hume took exactly the opposite direction—the scarcity of nature toward man.

In this destruction of the agrarian myth Hume was supported by Adam Smith, who claimed that: '*the savage nations of hunters and fishers are so miserably poor that they have to reduce their needs*' [25a]. Self-sufficient economy meant lack of cooperation and lack of exchange, and in fact it reduces man to an animal. The difference between animal and man lies in a special human feature, his innate propensity for exchange. We read in the famous *Wealth of Nations* the following: '*By nature a philosopher is not in genius and disposition half so different from a streetporter, as mastiff is from a greyhound, or a greyhound from a spaniel, or this last from a shepherd's dog*' [25b]. This impressive list provided him with arguments for the sake of exchange as the main human attribute. Although dogs are very different from each other, they do not exchange their services. In human society, the labour division meant that the country produced *means of subsistence* and the town produced *means of convenience and luxury* [25c]. The volume of the means of convenience and luxury shows the level of wealth of a given nation. Producing more commodities means the growth of the wealth of nations, which means more power for these nations and what means more inequality within them [25d]. Smith said: '*The first period of society, that of hunters, admits of no such inequality. Universal poverty establishes their universal equality, and the superiority either of age or of personal qualities*' [25e]. Smith's successor David Ricardo separated wealth from value. The rise of the commodity value is not connected with the rise of wealth. Wealth is quantity and utility, and the value depends on labour. Nature gives wealth but not value. We cannot get value for nothing [19a].

The Enlightenment replaced the old fertility and plenty with 'value'—a new category invented by a new science, 'Political Economy'. Political economy recognized the complexity of an old meaning of plenty (fertility) and said it was an incorrect mixture of wealth and value. As a result of this distinction, an idea appeared—we should not pay so much attention to things given for nothing. The political economists' curiosity was attracted to the idea of scarcity. Food gradually stopped being a main commodity. The controversy surrounding luxury faded away. The conditions for overcoming scarcity became the subject of reflection. Writers agreed that the modern understanding fertility as scarcity deepened social inequality, however they said that the alternative is poverty. Inequality became the requisite price we have to pay for our welfare. The labour division, specialisation and competition were identified as the main conditions of welfare. Smith was more an optimist. Ricardo and Malthus were not. They saw the main threat of the old economic base—agriculture. It turned out that the possibilities of industry were boundless, while the agricultural possibilities were limited [19b] [12] [13] [9]. It was the end of the Mother-Earth myth. The new commercial society needed a new secular fertility.

THE COMING AGE OF A NEW PLENTY?

People believed that they could produce this 'fertility' by themselves. They believed that such an understanding of welfare was all they needed. And since Lionel Robbins' very suggestive definition [20] [21], economics has been understood as the science which deals with the allocation of scarce resources for different uses. Scarcity became the main attribute of economic fact.

The alternative for those seeking a new fertility became socialism and Marxism. They wanted to save the old subject of the agrarian myth by returning to the problem of equality. As result of a gigantic social experiment, it became evident that we could share and share alike only while that division was the division of poverty, not of welfare.

The discussion about fertility and scarcity led to a replacement of the Mother-Earth myth with the mundane myth of the market. Philosopher Leszek Kołakowski said: '*seeking a myth is the affirmation of human needs: first, the need of significance and the need of faith in constancy of human values*'[8]. In this way, seeking myth is an expression of human disagreement with the fortuitousness of the world. The agrarian myth sets people in the conviction that nature is not the only thing that matters, that there is a plan of God too—Providence. From this belief the Marxist economy developed and man tried to be God. The market myth creates a situation in which man is afraid of his future, and he is disappointed with the market pricing his values. It does not make any difference what the main object of economy and economics is—fertility or scarcity. The Earth has made a long journey from the big mother to a cause of economic problems; from being the god to being the land, which gives his owner a financial bonus for non-cultivation.

The main feature of the modern economy is globalisation, which is understood as an economic balance within which there is no one sole economic power. Information is considered to be a main commodity. There is a new phenomenon of wide distribution of this commodity good free of charge (for example, the growth of Internet). Are these two phenomena the symptoms of the coming age of a new plenty?

Janina Rosicka
Jagiellonian University,
Chair of the History of Economic Thought,
Myslenice, Poland
e-mail: uurosick@cyf-kr.edu.pl

REFERENCES

[1a] **Eliade, M.,** *Traktat o historii religii*, introd. Kołakowski, L., Opus, 1993 (i ed. Traite d'histoire des religions, 1949), ch. 7, 8, 9.
[1b] *ibid.*, ch. 7, 9.
[2] **Fourier, Ch.,***Design for Utopia: Selected Writings of Charles Fourier*, Manuel, F.E., (foreword), Schocken Books, 1971.
[3a] **Guriewicz, A.,** *Kategorie kultury średniowiecznej*, Warszawa, 1976, pp. 221-222.
[3b] *ibid.*, pp. 216-293.
[3c] *ibid.*, p. 232.
[4] **Guriewicz, A.**, *Problemy średniowiecznej kultury ludowej*, PIW, Warszawa 1987, pp. 262-312.
[5] **Herskovits, M.J.**, *Cultural Anthropology*, New York, 1955, p. 164.
[6a] **Hume, D.**, "A Treatise of Human Nature", in: **Aiken H.D.** (ed.), *Hume's Moral and Political Philosophy*, The Hafner Library of Classics, Hafner Press, New York 10 ed., (i ed. 1749-30), p. 55.
[6b] *ibid.*, p. 63.
[7] **Jaspers, K.**, "Rozum i egzystencja", in: **Piecuch, C.,** (ed.),*Rozum i Egzystencja. Nietzsche a chrześcijaństwo*, Wydawnictwo Naukowe PWN, 1991 (i ed. Vernunft und Existenz, 1960), pp. 3-42 (the first lecture).
[8] **Kołakowski, L.**, *Obecność mitu*, Wydawnictwo Dolnoślaskie, Wrocław, 1994, p. 119.

[9] **LeMahieu, D L.**, "Malthus and The Theology Of Scarcity", *Journal of the History of Ideas*, **40**, pp. 467-74.
[10] **Locke, J.**, *Two Treatises of Goverment*, **Laslett, P.**, (ed.), Cambridge University Press, Cambridge, 1960 (i ed. 1690), pp. 46-48.
[11] **Machiavelli, N.**, *The Prince*, Oxford University Press, World Classics, Oxford, 1935 (i ed. 1532), ch. 16.
[12] **Malthus, T.R.**, "An Essay on the Principle of Population", in: *The Works of Thomas Robert Malthus*, **Wrigley, E. and Souden, D.** (eds.), William Pickering, London, **vol. 1-3**, 1986 (i ed. 1803).
[13] **Malthus, T.R.**, "An Inquiry into the Nature and Progress of Rent", in: **Wrigley, E. and Souden, D.** (eds.), *The Works of Thomas Robert Malthus*, William Pickering, London, vol.7, 1986 (i ed. 1815).
[14] **Marx, K.**, *Capital*, Lawrence & Wishart, London, 1970.
[15] **Marx, K.**, *A Contribution to the Critique of Political Economy*, Lawrence & Wishart, London, 1970.
[16] **Marx, K.**, *Theories of Surplus Value*, Lawrence & Wishart, London, 1972.
[17] **Proudhon, P.J.**, *Systeme des contradictions économiques ou philosophie de la misère*, Paris, 1846.
[18] **Quesnay, F.**, *François Quesnay et la Physiocratie*, Institut National d'Etudes Démographiques, Paris, 1958, ii vol.
[19a] **Ricardo, D.**, "On the Principles of Political Economy and Taxation", in: **Sraffa, P.** (ed.), *The Works and Correspondence*, 1 vol., Cambridge University Press, London, 1951 (i ed. 1817), ch. 20.
[19b] *ibid.*, ch. 32.
[20] **Robbins, L.**, "Economics and Political Economy", *American Economic Review*, **71**, 2, 1981.
[21] **Robbins, L.**, *Essay on Nature and Significance of Economic Science*, MacMillan, London, 1932.
[22] **Rosicka, J.**, "Myth and Science—Francois Quesnay", in: **Rosicka, J.** (ed.), *Physiocracy Yesterday and Today. Economy—Philosophy—Politics*, Cracow Academy of Economics, Cracow, 1996.
[23] **Rosicka, J.**, *O wyobraźni ekonomicznej Polaków*, Universitas, Kraków, 1991, pp. 9-62 (the replacement of pagan fertility by the order of God).
[24] **Saint-Simon, C.H.**, *Social Organization, the Science of Man, and Other Writings*, **Markham, F.** (ed.), Harper & Row, Torchbooks, New York, 1964.
[25a] **Smith, A.**, *Inquiry into the Nature and Causes of Wealth of Nations*, introd. Bax, E. B., George Bell and Sons, London, 1899 (i ed. 1776), v. i, p. 2.
[25b] *ibid.*, v. i, p. 17.

[25c] *ibid.*, v. i, p. 384.
[25d] *ibid.*, v. i, b. iii, ch. i.
[25e] *ibid.*, v. ii, p. 230.
[26] **Thomas, K.**, *Religion and Decline of Magic*, Penguin Books, 1988 (i ed. 1971), pp. 767-800.
[27] **Vickrey, W.S.**, "An Exchange of Questions between Economics and Philosophy", (i ed. 1953), in: **Phelps E.S.** (ed.), *Economic Justice. Selected Readings*, Penguin, London, 1973, p. 35.

9 780792 357612

MIX
Papier aus verantwortungsvollen Quellen
Paper from responsible sources
FSC® C105338

FSC
www.fsc.org

If you have any concerns about our products,
you can contact us on
ProductSafety@springernature.com

In case Publisher is established outside the EU,
the EU authorized representative is:
Springer Nature Customer Service Center GmbH
Europaplatz 3, 69115 Heidelberg, Germany

Printed by Libri Plureos GmbH
in Hamburg, Germany